数控车铣加工
初级（中英双语版）

赵 慧 史清卫 主 编
王 称 马绪鹏 王培磊 王启祥 副主编

清华大学出版社
北京

内 容 简 介

本书作为"数控车铣加工"系列教材初级分册,内容涵盖数控加工工艺、编程基础和机床操作三部分。本书教学内容紧密对接教育部发布的"1+X"数控车铣加工职业技能等级标准,重点介绍数控车床、数控铣床和加工中心的编程指令、夹具、刀具、工艺流程、设备操作等专业知识,根据技能型人才培养需求,科学设计了典型教学案例及配套教学资源,通过渐进式的任务学习及训练,学生可掌握数控编程及加工的基本方法、工艺常识和操作技能。本系列教材及配套资源可用于机械及相关专业的本科、专科、高职和专业培训院校的数控技术等课程的教材。

本书根据高职高专的教学特点,结合高职高专学生的实际学习能力和教学培养目标编写,可作为高职高专机械加工专业或其他相关专业的通用教材,也可作为成人教育院校的培训教材,还可供从事机械加工的工程技术人员参考。

本书封面贴有清华大学出版社防伪标签,无标签者不得销售。
版权所有,侵权必究。举报: 010-62782989, beiqinquan@tup.tsinghua.edu.cn。

图书在版编目(CIP)数据

数控车铣加工: 初级: 汉、英 / 赵慧,史清卫主编. —北京: 清华大学出版社,2024.5
ISBN 978-7-302-66283-9

Ⅰ. ①数… Ⅱ. ①赵… ②史… Ⅲ. ①数控机床-车床-加工工艺-职业技能-鉴定-教材-汉、英 ②数控机床-铣床-加工工艺-职业技能-鉴定-教材-汉、英 Ⅳ. ①TG519.1 ②TG547

中国国家版本馆 CIP 数据核字(2024)第 098086 号

责任编辑: 王 芳 薛 阳
封面设计: 刘 键
责任校对: 韩天竹
责任印制: 杨 艳

出版发行: 清华大学出版社
网　　址: https://www.tup.com.cn, https://www.wqxuetang.com
地　　址: 北京清华大学学研大厦 A 座　　邮　编: 100084
社 总 机: 010-83470000　　邮　购: 010-62786544
投稿与读者服务: 010-62776969, c-service@tup.tsinghua.edu.cn
质量反馈: 010-62772015, zhiliang@tup.tsinghua.edu.cn
印 装 者: 三河市君旺印务有限公司
经　　销: 全国新华书店
开　　本: 185mm×260mm　　印　张: 17.25　　字　数: 423 千字
版　　次: 2024 年 5 月第 1 版　　印　次: 2024 年 5 月第 1 次印刷
印　　数: 1~1500
定　　价: 59.00 元

产品编号: 098631-01

前　　言

随着自动化、数字化、网络化、智能化技术的快速发展及广泛应用,制造业的人才需求发生了很大变化,其需求对象由某一个领域单一技术的技能人才转变为"通才＋专才"复合型技术的技能人才。为进一步落实中国共产党第十九次全国代表大会提出的深化"产教融合"的重大任务,国务院在发布的《国家职业教育改革实施方案》中明确提出在职业院校、应用型本科高校启动"学历证书＋若干职业技能等级证书"制度(即"1＋X"职业技能等级证书制度)试点工作,明确了开展深度"产教融合""双元"育人的具体指导政策与要求。其中,"1＋X"职业技能等级证书制度是统筹考虑、全盘谋划职业教育发展、推动企业深度参与协同育人和深化复合型技术技能人才培养培训而做出的重大制度设计。

本书是"1＋X"职业技能等级证书(数控车铣加工)系列教材之一,是根据教育部数控技能型紧缺人才培养培训方案的指导思想,以及数控车铣加工职业技能等级证书的标准要求,结合当前数控技术的发展及教学规律编写而成的。本系列教材以数控车铣加工职业技能等级证书考核样题为基础,选用国内多种通用的CAD/CAM软件,从数控车和数控铣产品加工的典型任务入手,通过讲解工程图样及工艺文件,编制零件的数控加工工艺和加工程序,特别是针对数控车铣综合加工工艺进行案例分析,使学习者掌握数控机床加工编程、完成定位及联动加工、检测,并控制产品的加工精度、对数控机床的精度进行检验及排除数控机床的一般故障等技能。

目前,已经出版的数控车铣机床高职教材存在教学思路老套、教学内容更新慢、配套资源形式单一,以及教材贴合度差等弊端,且缺少相应资源,不满足当前的立体化、多媒体化、电子化、思政元素深度化等教学要求。针对这些现状,本书以最新的车铣设备作为编写硬件,采用EPIP教学模式,突出实践技能训练,运用深度递进式写作方法,是数控车铣教学的"交钥匙工程"项目,在图书市场上还未见到该类型出版物。本书出版后,在中职、高职学校及数控设备加工人员培训等市场上,会有很大的用户群体。

本书共分为6个项目。项目一和项目三由赵慧编写,项目二由史清卫编写,项目四由王称编写,项目五由马绪鹏编写,项目六由王培磊编写,王启祥参与了本书的翻译工作。

由于编者水平有限,书中难免会有不足,恳请使用本书的师生和读者批评指正。

编　者

2023年11月

Preface

With the rapid development and wide application of automation, digitization, networking and intelligent technologies, the demand for talents in the manufacturing industry has undergone great changes, that is, the skilled talents of a single technology in a certain field have been transformed into a composite type of "generalist + specialist". In order to further implement the major task of deepening the integration of production and education proposed by the 19th National People's Congress, the "National Vocational Education Reform Implementation Plan" issued by the State Council clearly proposes to start the pilot work of the "diploma certificate + several vocational skill grade certificates" system, i. e "1+X"vocational skill grade certificates system, in vocational colleges and application-oriented undergraduate colleges and universities, and the specific guiding policies and requirements for in-depth integration of production and education and "dual-element" education have been clarified. Among them, the 1+X vocational skill grade certificates system is a major system designed for overall consideration and planning for the development of vocational education, and promoting the deep participation of enterprises in collaborative education and deepening the training of compound technical and skilled personnel.

This book is one of the series of textbook for "1+X" vocational skill grade certificates (CNC turning and milling). It is compiled according to the guiding ideology of the training program of the Ministry of education for CNC skilled and urgently needed talents and the requirements of the vocational skill grade certificate for CNC turning and milling processing, combined with the current development of CNC technology and teaching rules. This series of textbooks is based on the examination sample questions of the vocational skill grade certificate of CNC turning and milling machining, using a variety of domestic general CAD/CAM software, starting from the typical tasks of CNC turning and milling, by understanding engineering drawings and technological documents, compiling CNC machining process and machining procedures of parts, especially for the case analysis of CNC turning and milling comprehensive machining technology, especially for the case analysis of CNC turning and milling comprehensive processing technology, so that learners can master CNC programming, positioning and linkage processing, detecting and controlling the machining accuracy of products, testing the accuracy of CNC machine tools, and eliminating general malfunctions of CNC machine tools, etc.

At present, the published higher vocational textbooks for CNC turning and milling have disadvantages such as old-fashioned teaching ideas, slow content update, supporting resources with single form and poor conformity with textbooks, and lack of ideological and political resources, which do not meet the current teaching requirements of stereoscopic, multimedia, computerization, and in-depth ideological and political elements. In response to these current situations, this textbook adopts the latest turning and milling equipment, the EPIP teaching mode, and in-depth progressive writing methods, highlights practical skills training. After the book is published, there will be a large user group in the secondary and higher vocational schools and the training of CNC equipment machining personnel.

This book contains 6 projects. Project 1 and project 3 is written by Zhao Hui. Project 2 is written by Shi Qingwei. Project 4 is written by Wang Chen. Project 5 is written by Ma Xupeng. Project 6 is written by Wang Peilei. Wang Qixiang Participates in the translation of this book.

Due to the limited level of editors, there will inevitably be flaws, deficiencies and errors in this book. I urge teachers, students and readers who use the book to criticize and correct me.

Editor
Nov. 2023

目 录

项目引导 ··· 1

项目一 销轴车削编程与加工训练 ·· 2

  任务一 学习关键知识点 ·· 3
    1.1 初步认识数控车床 ·· 3
    1.2 车削加工特点 ·· 5
    1.3 刀具知识 ·· 5
      1.3.1 刀具分类 ·· 5
      1.3.2 刀具选择 ·· 5
    1.4 夹具知识 ·· 6
    1.5 量具知识 ·· 7
      1.5.1 游标卡尺 ·· 7
      1.5.2 千分尺 ··· 9
    1.6 基本指令 ·· 10
      1.6.1 数控车削程序的基本格式 ··· 10
      1.6.2 基本准备功能指令 ·· 11
      1.6.3 辅助功能 M 代码 ·· 13
      1.6.4 数控车床利用 T 代码 ··· 14
      1.6.5 主轴功能 S 代码 ··· 14
  任务二 工艺准备 ·· 14
    1.7 零件图分析 ··· 14
    1.8 工艺设计 ·· 15
    1.9 数控加工程序编写 ··· 16
  任务三 上机训练 ·· 17
    1.10 设备与用具 ··· 17
    1.11 认识机床 ·· 17
      1.11.1 开机 ··· 17
      1.11.2 机床面板 ··· 18
      1.11.3 回参考点 ··· 23
    1.12 刀具准备 ·· 23
    1.13 设定工件原点 ·· 24
      1.13.1 Z 轴原点设定 ·· 24

1.13.2　X 轴原点设定 ………………………………………………… 24
1.14　程序编辑 ……………………………………………………………… 25
1.15　数控加工程序的仿真 ………………………………………………… 26
1.16　零件加工 ……………………………………………………………… 26
1.17　零件检测 ……………………………………………………………… 27
项目总结 …………………………………………………………………………… 28
课后习题 …………………………………………………………………………… 28

项目二　手柄车削编程与加工训练 …………………………………………… 31

任务一　学习关键知识点 …………………………………………………… 32
2.1　基本指令 ……………………………………………………………… 32
2.1.1　圆弧插补指令 ………………………………………………… 32
2.1.2　倒角过渡指令 ………………………………………………… 33
2.1.3　刀尖圆弧半径补偿指令 ……………………………………… 34
2.1.4　仿形粗加工循环指令 G73 …………………………………… 36

任务二　工艺准备 …………………………………………………………… 37
2.2　零件图分析 …………………………………………………………… 37
2.3　工艺设计 ……………………………………………………………… 37
2.4　数控加工程序编写 …………………………………………………… 38

任务三　上机训练 …………………………………………………………… 40
2.5　设备与用具 …………………………………………………………… 40
2.6　开机前检查 …………………………………………………………… 40
2.7　加工前准备 …………………………………………………………… 41
2.8　零件加工 ……………………………………………………………… 41
2.9　零件检测 ……………………………………………………………… 41

项目总结 …………………………………………………………………………… 42
课后习题 …………………………………………………………………………… 42

项目三　驱动轴车削编程与加工训练 ………………………………………… 44

任务一　学习关键知识点 …………………………………………………… 45
3.1　螺纹常识 ……………………………………………………………… 45
3.1.1　螺纹分类 ……………………………………………………… 45
3.1.2　公制螺纹的表示方法 ………………………………………… 45
3.1.3　螺纹车削相关尺寸计算 ……………………………………… 46
3.2　螺纹检测量具 ………………………………………………………… 46
3.3　基本指令 ……………………………………………………………… 46

任务二　工艺准备 …………………………………………………………… 48
3.4　零件图分析 …………………………………………………………… 48
3.5　工艺设计 ……………………………………………………………… 48

3.6　数控加工程序编写 …………………………………………………… 49
任务三　上机训练 ……………………………………………………………… 51
　　3.7　设备与用具 …………………………………………………………… 51
　　3.8　开机前检查 …………………………………………………………… 52
　　3.9　加工前准备 …………………………………………………………… 52
　　3.10　零件加工 ……………………………………………………………… 52
　　3.11　零件检测 ……………………………………………………………… 52
项目总结 …………………………………………………………………………… 53
课后习题 …………………………………………………………………………… 53

项目四　驱动轮铣削编程与加工训练 …………………………………………… 56

任务一　学习关键知识点 ……………………………………………………… 57
　　4.1　初步认识数控加工中心 ……………………………………………… 57
　　4.2　铣削加工特点 ………………………………………………………… 58
　　4.3　刀具知识 ……………………………………………………………… 59
　　4.4　夹具知识 ……………………………………………………………… 60
　　4.5　量具知识 ……………………………………………………………… 61
　　4.6　工艺知识 ……………………………………………………………… 62
　　　　4.6.1　定位原理 …………………………………………………… 62
　　　　4.6.2　基准 ………………………………………………………… 63
　　　　4.6.3　机械加工工艺规程制定的原则 …………………………… 64
　　4.7　基本指令 ……………………………………………………………… 64
　　　　4.7.1　数控铣削程序的基本格式 ………………………………… 64
　　　　4.7.2　常用准备功能指令 ………………………………………… 66
　　　　4.7.3　常用辅助功能指令 ………………………………………… 70
任务二　工艺准备 ……………………………………………………………… 71
　　4.8　零件图分析 …………………………………………………………… 71
　　4.9　工艺设计 ……………………………………………………………… 72
　　4.10　数控加工程序编写 …………………………………………………… 73
任务三　上机训练 ……………………………………………………………… 76
　　4.11　设备与用具 …………………………………………………………… 76
　　4.12　认识机床 ……………………………………………………………… 76
　　　　4.12.1　开机检查 …………………………………………………… 76
　　　　4.12.2　机床面板 …………………………………………………… 77
　　　　4.12.3　回原点 ……………………………………………………… 81
　　4.13　刀具准备 ……………………………………………………………… 81
　　4.14　寻边器与工件找正 …………………………………………………… 82
　　　　4.14.1　安装寻边器 ………………………………………………… 82
　　　　4.14.2　加工中心调用寻边器 ……………………………………… 83

4.14.3　工件找正 ··· 83
4.15　对刀 ··· 84
4.16　程序编辑 ··· 85
4.17　数控加工程序的仿真 ··· 86
4.18　零件加工 ··· 86
4.19　零件检测 ··· 87
项目总结 ·· 88
课后习题 ·· 88

项目五　槽轮铣削编程与加工训练 ·· 90

任务一　学习关键知识点 ·· 91
5.1　数控机床坐标系旋转功能及原理 ·· 91
5.2　子程序的格式与应用 ·· 93
5.2.1　子程序的格式 ··· 93
5.2.2　子程序的调用 ··· 93
5.2.3　子程序在相似形状重复加工时的应用 ·························· 94
5.2.4　子程序在分层加工时的应用 ···································· 95

任务二　工艺准备 ·· 96
5.3　零件图分析 ··· 96
5.4　工艺设计 ·· 97
5.5　数控加工程序编写 ·· 98

任务三　上机训练 ·· 101
5.6　设备与用具 ·· 101
5.7　开机检查 ··· 101
5.8　零件加工 ··· 101
5.9　零件检测 ··· 102

项目总结 ·· 102
课后习题 ·· 102

项目六　驱动轴、从动轴螺纹孔编程与加工训练 ······························ 105

任务一　学习关键知识点 ·· 106
6.1　常见丝锥 ··· 106
6.2　攻丝加工工艺方法 ·· 106
6.3　攻丝加工螺纹底孔尺寸 ··· 107
6.4　螺纹攻丝的操作方法 ·· 107
6.4.1　手工攻丝 ··· 107
6.4.2　机械攻丝 ··· 108
6.5　攻丝指令 ··· 109
6.6　螺纹孔的检测 ·· 110

任务二　工艺准备 …………………………………………………………………… 110
　　6.7　零件图分析 ………………………………………………………………… 110
　　6.8　工艺设计 …………………………………………………………………… 111
　　6.9　数控加工程序编写 ………………………………………………………… 111
任务三　上机训练 …………………………………………………………………… 113
　　6.10　设备与用具 ……………………………………………………………… 113
　　6.11　开机检查 ………………………………………………………………… 113
　　6.12　零件加工 ………………………………………………………………… 114
　　6.13　零件检测 ………………………………………………………………… 114
项目总结 ……………………………………………………………………………… 114
课后习题 ……………………………………………………………………………… 115

Contents

Projects Guidance ·· 117

Project 1 Programming and Machining Training for Pin Shaft Turning ············· 119

   Task 1 Learn Key Knowledge Points ·· 120
      1.1 Preliminary understanding of CNC lathes ··· 120
      1.2 Turning features ·· 123
      1.3 Knowledge of cutters ··· 123
         1.3.1 Classification of cutters ·· 123
         1.3.2 Cutters selection ·· 124
      1.4 Knowledge of fixtures ·· 125
      1.5 Knowledge of measuring tools ·· 126
         1.5.1 Vernier caliper ··· 126
         1.5.2 Micrometers ·· 128
      1.6 Basic instructions ·· 130
         1.6.1 Basic format of the CNC turning program ································ 130
         1.6.2 Basic preparatory function instructions ····································· 131
         1.6.3 Miscellaneous function M-codes ··· 135
         1.6.4 CNC lathes utilize T-code ··· 135
         1.6.5 spindle function S-code ·· 135
   Task 2 Technological Preparation ·· 136
      1.7 Parts drawing analysis ··· 136
      1.8 Technological design ··· 136
      1.9 CNC machining programming ·· 138
   Task 3 Hands-on Training ·· 139
      1.10 Equipment and appliances ··· 139
      1.11 Get to know the machine tool ·· 139
         1.11.1 Power on ·· 139
         1.11.2 Operation panel ··· 140
         1.11.3 Return to reference point ·· 146
      1.12 Cutters preparation ·· 147
      1.13 Set the workpiece origin ·· 147
         1.13.1 Z-axis origin setting ·· 147

    1.13.2 X-axis origin setting …… 148
  1.14 Program editing …… 149
  1.15 Simulation of CNC machining program …… 150
  1.16 Part machining …… 150
  1.17 Part inspection …… 152
 Project Summary …… 152
 Exercises After Class …… 152

Project 2 Programming and Machining Training for Handle Turning …… 156

 Task 1 Learn Key Knowledge Points …… 157
  2.1 Basic instructions …… 157
    2.1.1 Circular interpolation instruction …… 157
    2.1.2 Chamfer/Corner arc transition instruction …… 158
    2.1.3 Tool radius compensation instructions …… 160
    2.1.4 Profile modeling roughing turning cycle instruction G73 …… 161
 Task 2 Technological Preparation …… 162
  2.2 Part drawing analysis …… 162
  2.3 Technological Design …… 163
  2.4 CNC machining programming …… 164
 Task 3 Hands-on Training …… 166
  2.5 Equipment and appliances …… 166
  2.6 Check before powering on …… 166
  2.7 Preparation before machining …… 166
  2.8 Part machining …… 167
  2.9 Part inspection …… 167
 Project Summary …… 168
 Exercises After Class …… 168

Project 3 Programming and Machining Training for Drive Shaft Turning …… 171

 Task 1 Learn Key Knowledge Points …… 172
  3.1 General knowledge of thread …… 172
    3.1.1 Classification of threads …… 172
    3.1.2 Representation of metric threads …… 172
    3.1.3 Calculation of dimensions related to thread turning …… 173
  3.2 Thread measuring tool …… 174
  3.3 Basic instruction …… 174
 Task 2 Technological Preparation …… 176
  3.4 Part drawing analysis …… 176
  3.5 Technological design …… 176

		3.6	CNC machining programming	177

Task 3 Hands-on Training ········ 180
   3.7 Equipment and appliances ········ 180
   3.8 Check before powering on ········ 180
   3.9 Preparation before machining ········ 180
   3.10 Part machining ········ 181
   3.11 Part inspection ········ 181
Project Summary ········ 182
Exercises After Class ········ 182

Project 4 Programming and Machining Training for Drive Wheel Milling ········ 185

Task 1 Learn Key Knowledge Points ········ 186
   4.1 Preliminary understanding of CNC machining center ········ 186
   4.2 Milling characteristics ········ 188
   4.3 Knowledge of cutters ········ 189
   4.4 Knowledge of fixture ········ 191
   4.5 Knowledge of measuring tools ········ 192
   4.6 Technology knowledge ········ 193
      4.6.1 Positioning principle ········ 193
      4.6.2 Benchmarks ········ 194
      4.6.3 Principles for the formulation of machining process specification ········ 196
   4.7 Basic instructions ········ 196
      4.7.1 Basic format of CNC milling program ········ 196
      4.7.2 Commonly used preparatory function instructions ········ 198
      4.7.3 Commonly used auxiliary function instructions ········ 204
Task 2 Technological Preparation ········ 204
   4.8 Part drawing analysis ········ 204
   4.9 Technological design ········ 205
   4.10 CNC machining program writing ········ 207
Task 3 Hands-on Training ········ 210
   4.11 Equipment and appliances ········ 210
   4.12 Get to know the machine tool ········ 210
      4.12.1 Power on inspection ········ 210
      4.12.2 Operation panel ········ 211
      4.12.3 Return to origin ········ 216
   4.13 Cutters preparation ········ 216
   4.14 Edge finder and workpiece alignment ········ 217
      4.14.1 Installing the edge finder ········ 217

　　　　4.14.2　The machining center calls the edge finder ……………… 218
　　　　4.14.3　Workpiece alignment ……………………………… 218
　　4.15　Cutter setting ……………………………………………… 219
　　4.16　Program editing …………………………………………… 220
　　4.17　Simulation of CNC machining program …………………… 222
　　4.18　Part machining ……………………………………………… 222
　　4.19　Part inspection ……………………………………………… 224
　Project Summary ………………………………………………………… 224
　Exercises After Class ……………………………………………………… 224

Project 5　Programming and Machining Training for Geneva Wheel Milling ……… 228

　Task 1　Learn Key Knowledge Points ………………………………… 229
　　5.1　Function and principle of coordinate system rotation ………… 229
　　5.2　Format and application of subprogram ………………………… 231
　　　　5.2.1　Format of subprogram ……………………………… 232
　　　　5.2.2　Call subprogram …………………………………… 232
　　　　5.2.3　Application of subprogram when a similar shape is
　　　　　　　processed repeatedly ………………………………… 233
　　　　5.2.4　Application of subprogram in layered processing …… 234
　Task 2　Technological Preparation ……………………………………… 235
　　5.3　Part drawing analysis …………………………………………… 235
　　5.4　Technological design …………………………………………… 236
　　5.5　CNC machining program writing ……………………………… 237
　Task 3　Hands-on Training ……………………………………………… 240
　　5.6　Equipment and Appliances …………………………………… 240
　　5.7　Check before powering on …………………………………… 240
　　5.8　Part machine …………………………………………………… 241
　　5.9　Part inspection ………………………………………………… 241
　Project Summary ………………………………………………………… 241
　Exercises After Class ……………………………………………………… 242

Project 6　Programming and Machining Training for Threaded Hole of Drive Shaft and Driven Shaft …………………………………………… 244

　Task 1　Learn Key Knowledge Points ………………………………… 245
　　6.1　Common taps …………………………………………………… 245
　　6.2　Tapping process method ……………………………………… 246
　　6.3　Thread bottom hole size for tapping cutting …………………… 247
　　6.4　Operation method of thread tapping …………………………… 247
　　　　6.4.1　Hand tapping ………………………………………… 247

    6.4.2 Mechanical tapping ··· 248
  6.5 Tapping instructions ··· 249
  6.6 Inspection of threaded holes ···································· 250
Task 2 Technological Preparation ·· 251
  6.7 Part drawing analysis ·· 251
  6.8 Technological design ·· 252
  6.9 CNC machining program writing ································ 252
Task 3 Hands-on Training ··· 254
  6.10 Equipment and appliances ······································ 254
  6.11 Check before powering on ······································ 254
  6.12 Part machine ··· 255
  6.13 Part inspection ·· 255
Project Summary ··· 255
Exercises After Class ·· 256

项目引导

目前,数控加工技术在国内外的加工制造业中已经得到广泛应用。数控加工产业的高速发展对数控编程与操作技术人员也提出了更高的要求。以实际工程项目为导引、以实践应用为导向,可以更好地培养学生科学探究和解决问题的能力。槽轮机构在机械行业中应用广泛,用于实现间歇动作,其具有机构简单、工作可靠和传动效率高等特点。本书以典型摆轮间歇运动机构(见图0-1)的相关零件为载体,以6个典型项目(见图0-2)为主线,重点学习数控车削和数控铣削的工艺制定、数控编程与操作的专业知识和操作技能。

图 0-1 典型摆轮间歇运动机构

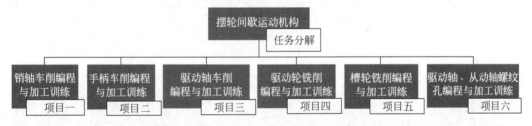

图 0-2 典型项目

摆轮间歇运动机构包括6个典型零件和若干标准件。针对这6个典型零件数控加工所需的专业知识和操作技能,分别进行学习和训练,完成零件的加工,并最终完成机构装配。每个项目都包括完成零件加工所必须的专业知识和技能的学习,如设备、夹具、刀具、基本指令和机床操作等。根据数控加工的基本流程,进行有针对性的学习和训练,以达到熟练掌握数控车削和铣削的加工工艺设计、数控程序编写和机床操作的目的。

项目一　销轴车削编程与加工训练

> **思维导图**

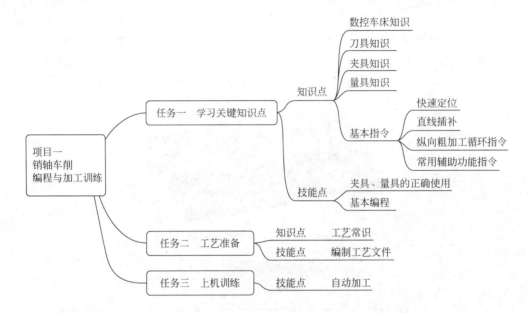

> **学习目标**

知识目标

(1) 具备轴类零件的识图能力。
(2) 了解阶梯轴的用途和特点。
(3) 理解纵向加工循环指令各参数的含义。

能力目标

(1) 掌握外圆车刀和切断刀的选择和安装方法。
(2) 能够独立确定加工工艺路线,并正确填写工艺文件。
(3) 能够正确操作数控车床,并根据加工情况调整加工参数。
(4) 能够根据零件结构特点和精度合理选用量具,并正确、规范地测量相关尺寸。

素养目标

(1) 培养学生的科学探究精神和态度。
(2) 培养学生的工程意识。
(3) 培养学生的团队合作能力。

项目一 销轴车削编程与加工训练

▶任务引入

阶梯轴最主要的作用就是定位安装的零件。其高低不同的轴肩可以限制轴上的零件延轴线方向的运动或运动趋势,防止安装的零件在工作中产生滑移,并能减小工作中一些零件产生的轴向压力对其他零件的影响。阶梯轴在生产及生活中应用广泛。

根据零件图(见图1-1)要求,制定加工工艺、编写数控加工程序,并完成拨销零件的加工。该零件作为典型的轴类零件,材料为45钢,毛坯调质处理,要求表面光整无划伤。

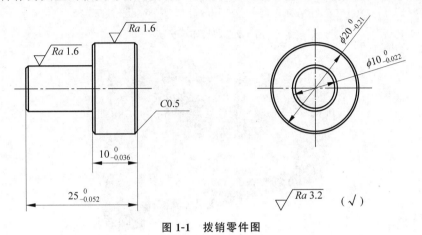

图1-1 拨销零件图

任务一 学习关键知识点

1.1 初步认识数控车床

数控车床是使用较为广泛的数控机床之一。它主要用于轴类零件或盘类零件的内、外圆柱面,任意锥角的内、外圆锥面,复杂回转内、外曲面和圆柱、圆锥螺纹等切削加工,并能进行切槽、钻孔、扩孔、铰孔及镗孔等加工工艺。

数控机床是按照事先编制好的加工程序,自动地对被加工零件进行加工。编程人员把零件的加工工艺路线、工艺参数、刀具的运动轨迹、位移量、切削参数及辅助功能,按照数控机床规定的指令代码及程序格式编写成加工程序单,再把这个加工程序单中的内容记录在控制介质上,然后输入数控机床的数控装置中,从而实现控制机床对零件的加工。

数控车床由数控装置、床身、主轴箱、刀架进给系统、尾座、液压系统、冷却系统、润滑系统、排屑器等部分组成。数控车床分为立式数控车床和卧式数控车床两类,如图1-2所示。立式数控车床用于回转直径较大的盘类零件车削加工;卧式数控车床用于轴向尺寸较长或小型盘类零件的车削加工。

CK6150e数控车床(见图1-3)采用平床身结构,主轴采用前、后端两点支撑的典型结构,具有很高的刚度;主传动为三挡无级变频调速,速度范围为45~1600r/min;主传动齿轮副均经淬硬磨削处理,各传动副和滚动轴承均经强力油液润滑,具有良好的高速低温升性能;主轴箱箱体的设计充分考虑了散热措施和减振机构,使主轴箱具有噪声低、传动精度高的特点。

(a) 立式数控车床　　　　　　　　(b) 卧式数控车床

图 1-2　数控车床分类

图 1-3　CK6150e 数控车床

该机床主轴由变频电机通过 V 带经变速机构驱动主轴箱，通过变频系统控制变频电机可实现 100～2000r/min（正反转）手动三挡无级调速。横向（X 轴）及纵向（Z 轴）进给运动均由伺服电机驱动精密滚珠实现。滑板导轨贴有防爬行的塑料软带，可很好地保证机床的定位精度和重复定位精度。机床配置手动三爪自定心卡盘和四工位数控刀架，具有定位精度高、稳定可靠、应用范围广、结构简单、维修方便等特点。

CK6150e 数控车床主要参数如表 1-1 所示。

表 1-1　CK6150e 数控车床主要参数

项 目 内 容	技 术 规 格
床身上最大回转直径/mm	ϕ500
滑板上最大回转直径/mm	ϕ300
两顶尖间最大工件长度/mm	880
主轴通孔直径/mm	ϕ80
主轴转速范围/(r/min)	100～2000
主轴电机功率/kW	7.5（变频）
数控刀架	立式四工位刀架
X/Z 轴行程/mm	310/880
X/Z 轴快移速度/(mm/min)	8000/10000
X/Z 轴进给速度范围/(mm/min)	1～2000
尾座套筒直径/mm	ϕ75
尾座套筒行程/mm	150

续表

项目内容	技术规格
尾座套筒锥孔锥度	莫氏5号
机床外形尺寸(长×宽×高)/(mm×mm×mm)	2900×1500×1800
机床净重/kg	2800
数控系统	FANUC 0i TF

1.2 车削加工特点

车削是以工件的旋转运动为主的运动,以刀架作进给运动的一种切削加工方法。其特点如下。

(1) 车削加工相对效率较高,其具有比铣削和磨削更高的效率。车削可以采用较高的工件转速和大切削深度(背吃刀量),从而实现高效加工。

(2) 车削加工生产效率高、加工范围广,可使用各种刀具完成内圆柱面、外圆柱面、端面、圆锥面、沟槽(直角沟槽、圆弧槽及异形槽等)和螺纹等的加工任务。

(3) 车削加工具有较高的加工精度,其经济加工精度一般为IT9~IT7,表面粗糙度 Ra 值一般为 $12.5 \sim 1.6 \mu m$。精密车削精度可达 IT5,表面粗糙度 Ra 值可达到 $0.20 \mu m$。

如图1-4所示,在车削加工中切削速度通常用主轴转速 n 来表示,是指工件在主轴上每分钟的转数,单位为 r/min (rpm);走刀速度通常用工件与刀具每转的相对移动路程 f_n 来表示,单位为 mm/r;切削深度可分为轴向切深和径向切削,均用 a_p 表示,单位为 mm。

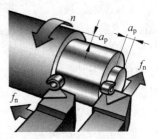

图1-4 车削加工模型

1.3 刀具知识

刀具对于数控车削加工来说至关重要,刀具的种类、材料和角度的选择直接影响加工零件的尺寸精度、表面质量和刀具的使用寿命。

1.3.1 刀具分类

根据制造刀具所用的材料可分为高速钢刀具、硬质合金刀具、金刚石刀具、立方氮化硼刀具、陶瓷刀具和涂层刀具等。

车削加工刀具种类很多,一般按结构可分为整体车刀、焊接车刀、机夹车刀和可转位车刀。按加工用途可分为外圆车刀、镗孔刀、切槽刀、外螺纹车刀等,如图1-5所示。

1.3.2 刀具选择

车刀的选择对于数控车削加工来说至关重要,其选用步骤如下。

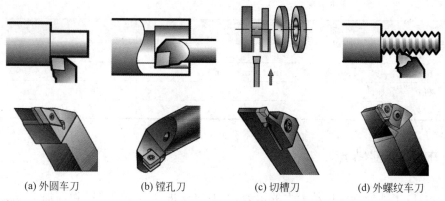

(a) 外圆车刀　　　　(b) 镗孔刀　　　　(c) 切槽刀　　　　(d) 外螺纹车刀

图 1-5　常用加工车刀

（1）确定加工类型。根据零件被加工部位的特征及加工策略，明确粗、精加工及内、外轮廓加工等加工类型（如外圆车削、端面车削、仿形车削及槽加工等）。

（2）选定刀具类型。根据工序内容选择刀具类型（如外圆车刀、镗孔刀、切槽刀及外螺纹车刀等）。

（3）确定刀杆尺寸。根据机床要求和孔径尺寸选择刀杆尺寸。

（4）选择刀片。根据轮廓特征及加工要求确定刀片形状、型号、槽型、刀尖圆弧半径及刀具牌号。数控车削加工多使用机夹可转位刀具，刀具的选择主要是刀片的选择，即刀片材料的选择、刀片尺寸的选择、刀片形状的选择、刀片刀尖圆弧半径的选择。刀片材料主要依据被加工工件的材料、工件表面的精度要求、切削载荷的大小及切削加工过程中有无冲击和振动等条件决定；刀片尺寸主要是指有效切削刃的长度，根据背吃刀量 a_p 和主偏角 k_r 确定，在使用刀片时可查阅相关手册；刀片形状根据被加工工件表面形状、切削方法、刀具寿命和刀片的可转位次数等因素来选择；刀片刀尖圆弧半径的大小直接影响刀尖的强度和被加工零件的表面粗糙度。通常在背吃刀量较小的精加工、细长轴加工或机床刚度较差的情况下，选取较小的刀片刀尖圆弧半径；在需要刀刃强度高、零件直径大的粗加工中，选取较大的刀片刀尖圆弧半径。

1.4　夹具知识

夹具是指在机械制造过程中用来固定加工对象，使加工对象占有正确的位置，以接受施工或检测的装置，又称卡具。从广义上说，在工艺过程的任何工序中，用以迅速、方便、安全地安装工件的装置，都可称为夹具。夹具通常由定位元件（确定工件在夹具中的正确位置）、夹紧装置、对刀引导元件（确定刀具与工件的相对位置或导引刀具方向）、分度装置（使工件在一次安装中能完成数个工位的加工，分为回转分度装置和直线移动分度装置两类）、连接元件及夹具体（夹具底座）等组成。

车床最常用的夹具为卡盘，主要用于固定加工零件的端面、外圆、内控及螺纹等各种成型面，是最常用的夹具之一。卡盘按驱动卡爪所用动力的不同分为，手动卡盘和动力卡盘两种。手动卡盘由卡盘体、活动卡爪和卡爪驱动机构组成。常用的手动卡盘有自动定心的三爪自定心卡盘和每个卡爪都可以单独移动的四爪卡盘。手动三爪自定心卡盘上三个卡爪导向部分的下部有螺纹,与碟形伞齿轮背面的平面螺纹相啮合,当用扳手通过四方孔转动小伞齿轮时,碟形齿轮转动,背面的平面螺纹同时带动 3 个卡爪向中心靠近或退出,用以夹紧不同直径的工件。也可使用反爪,用来安装直径较大的工件。动力卡盘主要有液压卡盘和气动卡盘。图 1-6 所示为数控车床常见的通用夹具。

　　(a) 三爪自定心卡盘　　　(b) 四爪卡盘　　　(c) 液压卡盘　　　(d) 气动卡盘

图 1-6　数控车床常见的通用夹具

1.5　量具知识

　　量具是实物量具的简称,是指在使用时具有固定形态、用以复现或提供给定量的一个或多个已知量值的器具。机械加工中常用的量具包括标准器具、通用器具及专用器具。标准器具是指用作测量或检定标准的量具,如量块、多面棱体、表面粗糙度比较样块等;通用器具也称万能量具,一般是指由量具厂统一制造的通用性量具,如直尺、平板、角度尺、卡尺和千分尺等;专用器具也称非标量具,是指专门为检测工件某一技术参数而设计制造的量具,如内外沟槽卡尺、钢丝绳卡尺、步距规等。量具是以固定形式复现量值的测量器具,其特点如下。

　　(1) 本身直接复现了单位量值,即量具的标称值就是单位量值的实际大小,例如,量块本身就复现了长度量的量值标准,即单位。

　　(2) 在结构上,一般没有测量机构,没有指示器或运动的元部件,如量块只是复现单位量值的一个实物。

　　(3) 由于没有测量机构,若不依赖其他配用的测量器具,则不能直接测出被测量值,例如,量块要配用干涉仪、光学计,因此,它是一种被动式测量器具。

1.5.1　游标卡尺

　　车削加工中最常用的量具为游标卡尺,如图 1-7 所示,是一种测量长度、内径、外径、深度的量具。游标卡尺由主尺和附在主尺上能滑动的游标尺两部分构成。主尺一般以毫米为单位,而游标尺上则有 10、20 或 50 个分格,根据分格的不同,游标卡尺可分为 10 分度游标卡尺、20 分度游标卡尺及 50 分度游标卡尺。10 分度的游标尺长度为 9mm,20 分度的游标尺长度为 19mm,50 分度的游标尺长度为 49mm。游标卡尺的主尺和游标尺上有两副活动量爪,分别是内测量爪和外测量爪,内测量爪通常用来测量内径,外测量爪通常用来测量长度和外径。

　　游标卡尺是在工业上常用的测量长度的仪器,它由主尺及能在主尺尺身上滑动的游标尺组成。从背面看,游标卡尺是一个整体。游标尺与主尺尺身之间有一弹簧片,利用弹簧片

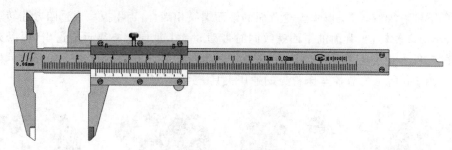

图 1-7 游标卡尺

的弹力使游标尺与主尺尺身靠紧。游标尺上部有一紧固螺钉,可将游标尺固定在主尺尺身上的任意位置。主尺尺身和游标尺都有量爪,利用内测量爪可以测量槽的宽度和管的内径,利用外测量爪可以测量零件的厚度和管的外径。深度尺与主尺尺身连在一起,可以测量槽和筒的深度。主尺尺身和游标尺上面都有刻度。以精确到 0.1mm 的游标卡尺为例,主尺尺身上的最小分度值是 1mm,游标尺上有 10 个小的等分刻度,总长度 9mm,每一分度值为 0.9mm,与主尺上的最小分度值相差 0.1mm。当两量爪并拢时,主尺尺身和游标尺的 0 刻度线对齐,它们的第 1 条刻度线相差 0.1mm,第 2 条刻度线相差 0.2mm……第 10 条刻度线相差 1mm,即游标尺的第 10 条刻度线恰好与主尺的 9mm 刻度线对齐。当两量爪间所测量的物体的长度为 0.1mm 时,游标尺应向右移动 0.1mm,这时它的第 1 条刻度线恰好与主尺尺身的 1mm 刻度线对齐。同样,当游标尺的第 5 条刻度线跟主尺尺身的 5mm 刻度线对齐时,说明两量爪之间有 0.5mm 的宽度,以此类推。当游标卡尺测量大于 1mm 的长度时,整的毫米数要从游标尺 0 刻度线与主尺尺身相对的刻度线读出。

游标卡尺在读数时,首先以游标尺 0 刻度线为准在主尺尺身上读取毫米整数,即以毫米为单位的整数部分。然后看游标尺上第几条刻度线与主尺尺身的刻度线对齐,如第 6 条刻度线与主尺尺身刻度线对齐,则小数部分即为 0.6mm(若没有正好对齐的刻度线,则取最接近对齐的刻度线进行读数)。如有零点误差,则一律用上述结果减去零点误差(若零点误差为负,则相当于加上相同大小的零点误差)。因此,读数结果为

$$L = 整数部分 + 小数部分 - 零点误差$$

判断游标尺上哪条刻度线与主尺尺身刻度线对准,可用下述方法:选定游标尺上相邻的三条刻度线,若左侧的刻度线在主尺尺身对应刻度线的右侧,右侧的刻度线在主尺尺身对应刻度线的左侧,则中间那条刻度线便可以认为是对准了,此时

$$L = 对准前刻度 + 游标尺上第\ n\ 条刻度线与主尺尺身的刻度线对齐 \times 分度值$$

如需测量几次取平均值,则不需要每次都减去零点误差,只要从最后结果中减去零点误差即可。

下面以图 1-8 所示精度为 0.02mm 的游标卡尺的某一状态为例进行说明。

(1) 在主尺上读出游标尺 0 刻度线以左的刻度,该值就是最后读数的整数部分。图 1-8 所示为 33mm。

(2) 游标尺上一定有一条刻度线与主尺的刻度线对齐,在游标尺上读出该刻度线距游标尺的 0 刻度线以左的刻度的格数为 12,乘以该游标卡尺的精度 0.02mm,就得到最后读数的小数部分。或者直接在游标尺上读出该刻度线的读数,图 1-8 所示为 0.24mm。

(3) 将所得到的整数部分和小数部分相加,就得到总尺寸为 33.24mm。

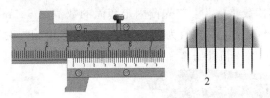

图 1-8 游标卡尺读数

1.5.2 千分尺

千分尺(螺旋测微器)作为数控车削零件测量的通用量具,应用非常广泛。其按照显示方式分为机械式千分尺和数显式千分尺,如图 1-9 所示。按照用途分为外测千分尺、内测千分尺、深度千分尺、螺纹千分尺、壁厚千分尺等。

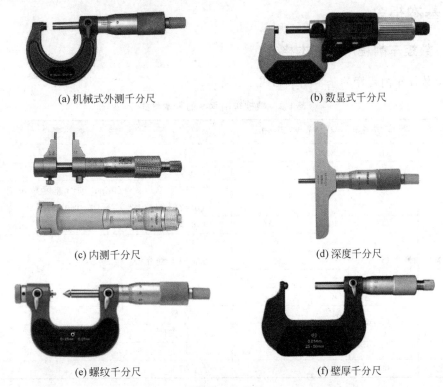

(a) 机械式外测千分尺　　(b) 数显式千分尺

(c) 内测千分尺　　(d) 深度千分尺

(e) 螺纹千分尺　　(f) 壁厚千分尺

图 1-9 千分尺

其中,外测千分尺,尤其是机械式外测千分尺最为常用。其主要由尺身、固定测头、测微螺杆、固定套筒、微分筒和测力棘轮组成,如图 1-10 所示。机械式外测千分尺具有成本低、测量准确、使用稳定可靠等特点,主要用于测量零件的外径和长度。

在利用千分尺测量零件的长度时,应一只手控制千分尺尺身的稳定性,另一只手转动测力棘轮,使得千分尺的测量端与零件的被测量部位充分贴合。读数时分为三个步骤:首先以微分筒的左侧端面为基准线,读固定套筒上刻度线的分度值(只读整数);再以固定套筒上的水平横线作为基准线,读微分筒上的刻度值(每格 0.01mm);最后再根据微分筒的实际位置估计尺寸的千分位(0.001mm)数值,若刻度线与基准线对齐则该尺寸的千分位为"0"。然后将三个数值相加,即为最终的尺寸数值。

如图 1-11 所示，其读数步骤为：首先读固定套筒的数值为 5.5mm，然后读微分筒的数值为 0.04mm，最后估计千分位的数值为 0.003mm，三个数值相加的结果为 5.543mm。

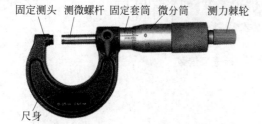

图 1-10　机械式外测千分尺结构

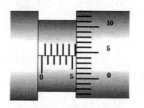

图 1-11　千分尺读数

1.6　基本指令

1.6.1　数控车削程序的基本格式

一个数控车削程序的基本格式如表 1-2 所示。

表 1-2　数控车削程序的基本格式

程序内容	程序步骤	程序语句	注　解
程序头	①	O0001;	程序名
程序体	②	T0101;	调用刀具及刀具补偿
	③	G97 G99 S800 M03;	主轴转速单位为 r/min；进给速度单位为 mm/r；主轴正转启动，转速为 800r/min
	④	G0 X__ Z__ M8;	快速移动到点（X__ Z__），并打开冷却液
	⑤	... ...	切削加工程序段
	⑥	G0 X__ Z__;	退回到安全位置
	⑦	M5;	主轴停转
	⑧	M9;	冷却液关闭
	⑨	...	如需其他刀具加工，则重复步骤②～⑧
程序尾	⑩	M30;	程序结束

程序头一般包括程序起始符和程序名或程序号，程序起始符根据不同的数控系统，可以为"％"":""MP"等不同的标识，有的系统也可以省略，在书写程序时需参照机床编程说明书进行指定。在 FANUC 0i 系统中，程序头可以直接以英文大写字母 O 后接数字的程序号来表示，程序号为正整数，取值范围为 0001～9999。通常 O9000～O9999 范围内的程序被锁定，用于保存机床特定功能的程序。在输入程序号时，数字前的 0 可以省略。程序由若干个程序段组成，每个程序段后以"；"结束。

程序体为程序的主体部分，为刀具切削工件的过程。加工过程若只有一把刀，则程序体为步骤②～⑧；若有多把刀具，则重复步骤②～⑧过程。其中，步骤②～④为切削过程的初始化，在这些步骤中调用刀具及刀具补偿，设置主轴转速及进给速度单位，并启动主轴（注意主轴旋转方向）。然后，控制刀具快速运动到安全位置，为正式切削加工做准备。步骤⑤为正式的切削过程,表示工件上一个区域的加工过程。若一个区域需要多次走刀切削,则只需重复"进刀、切削、退刀"的过程即可;若切削多个区域,只需重复步骤⑤~⑥即可。在数控车床加工过程中,每把刀具在所有区域切削完成后,应将刀具退回到安全位置,以便零件测量和装卸,以及防止发生碰撞。最后,关闭主轴和冷却液。

程序尾包括程序的结束指令和结束标识符。FANUC系统常用的程序结束指令为M02、M30(通常使用M30)。与程序头相似,结束标识符应根据机床编程说明书中的要求进行指定。

FANUC系统常用准备功能指令如表1-3所示。

表1-3 FANUC系统常用准备功能指令表

指令	用途	指令	用途
G00	快速定位	G42	刀尖圆弧半径右补偿
G01	直线插补	G70	精加工循环指令
G02	顺时针圆弧插补	G71	纵向粗加工循环
G03	逆时针圆弧插补	G72	横向粗加工循环
G04	暂停	G73	仿形粗加工循环
G20	英制输入	G92	螺纹加工循环
G21	公制输入	G96	恒线速度控制
G28	返回参考点	G97	恒转速控制
G40	取消刀尖圆弧半径补偿	G98	每分钟进给速度
G41	刀尖圆弧半径左补偿	G99	每转进给量

1.6.2 基本准备功能指令

1. 快速定位指令 G00

G00指令是指将刀具从当前点,以数控系统预设的最大进给速度(实际运行速度需考虑快速倍率旋钮的位置),快速移动到程序段所指令的下一个定位点。如图1-12所示,快速定位过程可能存在多种运动轨迹,常见的轨迹为由点 A 先定位到点 C,再由点 C 运动到点 B。由于机床在出厂时参数设定不同,轨迹也不同,在操作机床时应注意其运动轨迹,以免发生碰撞事故。快速定位指令格式为指令代码后接定位终点位置坐标。

指定刀具的移动有两种方法,即绝对坐标指令和相对(增量)坐标指令。绝对坐标指令是对刀具的移动位置以工件坐标系实际坐标值的方法指定终点坐标。在编程时,径向和轴向分别用地址 X__ 和 Z__ 表示。相对坐标又称增量坐标指令,是以刀具相对当前点以增量的形式指定终点坐标的方法,径向用地址 U__ 表示,轴向用地址 W__ 表示。绝对坐标和增量坐标在同一程序段中可以混合使用。

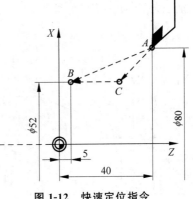

图1-12 快速定位指令

注意:在数控车削编程时,为了保持标注、编程和测量尺寸的一致性,径向坐标通常用直径值表示,即直径编程。虽然通过修改机床参数也可以用半径值表示,但很少使用。

指令格式为：

```
G00 X__ Z__;
```

例如，以下 4 句指令均可实现由点 A 快速移动到点 B：

```
G00 X52 Z5;
G00 U-28 W-35;
G00 X52 W-25;
G00 U-28 Z5;
```

建议初学者尽量使用绝对坐标指令，以减少不必要的失误。

2. 直线插补指令 G01

G01 指令是指将刀具以直线形式按 F 代码指定的速率，从其当前位置移动到命令要求的位置，如图 1-13 所示。其特点是各坐标以联动的方式，按进给速度 F 作任意斜率的直线运动。其指令格式为：

```
G01 X__ Z__ F__;
```

例如，由点 A 以指定的速度 F 移动到点 B 的指令为：

```
G01 X52 Z5 F0.2;
```

或

```
G01 U-28 W-35 F0.2;
```

3. 纵向粗加工循环指令 G71

G71 指令只需指定粗加工背吃刀量、退刀量、加工余量和精加工路线，系统将自动给出粗加工路线和加工次数，完成内、外圆表面的粗加工。如图 1-14 所示，G71 指令格式为：

```
G71 U (Δd) R (e);
G71 P (ns) Q (nf) U (Δu) W (Δw) F(f);
```

其中，Δd 表示每次的背吃刀量，一般用半径值指定，如 45 钢件取 1～2mm，铝件取 1.5～3mm；e 表示每次 X 轴方向退刀量，用半径值指定，一般取 0.5～1mm；ns 表示精加工轮廓程序段中的开始程序段号；nf 表示精加工轮廓程序段中的结束程序段号；Δu 表示 X 轴方向精加工余量，在加工内轮廓时，为负值；Δw 表示 Z 轴方向精加工余量，一般取 0.05～0.1mm。

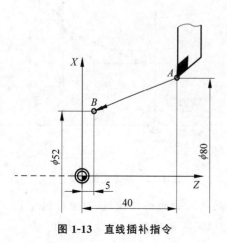

图 1-13 直线插补指令

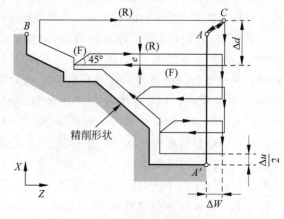

图 1-14 纵向粗加工循环指令

指令动作：从循环起点快速进刀，切削加工，45°退刀，快速返回。进刀，切削加工，45°退刀，快速返回。以此类推，完成粗加工。

注意：使用 G71 指令粗加工时，包含在 ns～nf 的程序段中的 F、S 代码对粗车循环无效。顺序号为 ns～nf 的程序段中不能调用子程序。G71 指令主要用于 X 轴、Z 轴方向同时单调增大或单调减少的轮廓加工。通常，ns～nf 的程序段中第一句必须沿 X 轴方向运动（FANUC 0i-mate 系统）。

4. 精加工循环指令 G70

G70 指令的功能是去除精加工余量，其指令格式为：

G70 P(ns) Q(nf) F(f);

G70 指令一般用于精加工，切除 G71 指令粗加工后留下的加工余量。

注意：在 ns～nf 的程序段中的 F、S 指令有效；在 G70 指令切削后，刀具回到循环起点。

5. 主轴转速控制指令

主轴转速控制指令包括两个指令：恒线速度控制指令 G96 和恒转速控制指令 G97。

G96 指令用于指定主轴以恒线速度旋转，主要用于加工径向尺寸变化大的外形轮廓，可使被加工部位的表面质量趋于一致，即主轴转速随着切削直径的变化而发生改变。由公式 $n=\dfrac{1000D}{\pi d}$ 可知，刀具越靠近中心，主轴转速越高。为防止主轴转速持续升高，通常需要与限制主轴最高转速指令 G50 配合使用。例如，设定主轴旋转的线速度为 120m/min，最高转速不超过 2000r/min，程序为：

G50 S2000;
G96 S120;

G97 指令用于指定主轴以恒转速的方式旋转，应用最为广泛。例如，设定主轴转速为 960r/min，方向正转，程序为：

G97 S960 M3;

6. 切削速度控制指令

切削速度控制指令包括以下两种。

(1) 每分钟进给速度指令 G98：用于设定刀具进给速度，单位为 mm/min。

(2) 每转进给量指令 G99：用于设定刀具进给量，单位为 mm/r。

两者之间的关系为

$$F_m = F_r \cdot n$$

其中，F_m 为每分钟进给速度；F_r 为每转进给量；n 为主轴转速。

注意：机床实际运动速度的大小还与进给倍率有关。

1.6.3 辅助功能 M 代码

辅助功能 M 代码用于控制机床的一些辅助动作，如主轴启停和冷却液开关等，FANUC 系统常用辅助功能 M 指令如表 1-4 所示。

表 1-4 FANUC 系统常用辅助功能指令表

序号	指令	功能	备注
1	M00	程序暂停	
2	M01	选择性暂停	配合机床开关控制程序暂停
3	M02	程序结束	
4	M03	主轴正转	
5	M04	主轴反转	
6	M05	主轴停转	
7	M08	冷却液打开	
8	M09	冷却液关闭	
9	M30	程序结束,并返回程序头	

注:通常每个程序段只允许写一个 M 代码。

1.6.4 数控车床利用 T 代码

数控车床利用 T 代码用于调用刀具和刀具补偿,指令格式为:

T__

刀具功能指令由字母 T 和 4 位数字组成,前两位数字代表刀具号,后两位数字代表刀具补偿号,在通常情况下,刀具号与刀具补偿号相同。

1.6.5 主轴功能 S 代码

数控车床使用主轴功能 S 代码指定机床主轴的回转速度,单位为 r/min 或 m/min,当主轴功能 S 代码指定的数值为正整数,且数值范围大于机床的最大额定转速时,机床主轴以最大额定转速旋转。指令格式为:

S__

辅助功能 M 代码控制主轴的启停,M03 为主轴正转、M04 为主轴反转、M05 为主轴停转,逆着 Z 轴正方向观察,主轴逆时针旋转定义为正转。

例如,指定主轴转速为 800r/min,正转,其指令格式为:

S800 M3;

任务二 工艺准备

1.7 零件图分析

根据零件的使用要求,选择 45 钢作为本零件的毛坯材料,毛坯下料尺寸定为 $\phi 25 \times 60$。在加工时,首先以 $\phi 25$ 毛坯外圆作为粗基准,粗加工 $\phi 10$ 和 $\phi 20$ 阶梯轴端面及外圆,然后精加工,并保证尺寸精度及表面质量,最后在切断零件时,保证长度尺寸的要求。

该阶梯轴的两个外圆需要在一次装夹的情况下,同时完成两个圆柱面的加工,以保证较好的同轴度。如果分为两次定位,则零件的校正难度较大,不容易保证零件的工作需要。

注意:在装夹毛坯时,应注意棒料伸出的长度,以免刀具与卡盘发生碰撞。

1.8 工艺设计

根据零件图分析,确定工艺过程,如表1-5所示。

表1-5 工艺过程卡片

机械加工 工艺过程卡片	产品型号	CLJG-01	零部件序号	BX-01	第1页	
	产品名称	槽轮机构	零部件名称	拨销	共1页	
材料 牌号	C45	毛坯规格	$\phi25\times60$	毛坯质量	kg	数量	1

工序号	工序名	工序内容	工段	工艺装备	工时/min	
					准结	单件
5	备料	按$\phi25\times60$尺寸备料	外购	锯床		
10	车加工	以$\phi25$毛坯外圆作为粗基准,精加工$\phi10$和$\phi20$外圆表面及端面	车	车床外径千分尺	45	30
15	清理	清理工件,锐角倒钝	钳			5
20	检验	检验工件尺寸	检			5

本训练任务针对第10工序车削加工,进行工序设计,制订工序卡片,如表1-6所示。

表1-6 车削加工工序卡片

机械加工 工序卡片	产品型号	CLJG-01	零部件序号	BX-01	第1页
	产品名称	槽轮机构	零部件名称	拨销	共1页

工序号	20
工序名	车加工
材料	C45
设备	数控车床
设备型号	CK6150e
夹具	三爪自定心卡盘
量具	游标卡尺
	千分尺
准结工时	45min
单件工时	30min

工步	工步内容	刀具	S/ (r/min)	F/ (mm/r)	a_p/ mm	工步工时/min	
						机动	辅助
1	工件安装						5
2	粗加工$\phi10$和$\phi20$外圆表面及端面,精加工余量为0.2mm	外圆粗车刀	1200	0.2	1.5	15	
3	精加工$\phi10$和$\phi20$外圆表面及端面	外圆精车刀	1500	0.1	0.2	10	
4	切断并倒角	切断刀(刀宽3mm)	700	0.07	0.1	10	
5	拆卸、清理工件						5

1.9 数控加工程序编写

根据工序加工工艺,编写加工程序,如表1-7所示。

表1-7 拨销数控加工程序

序号	程序语句	注　解
	O0001;	
N1	T0101;	调用外圆车刀
	G97 G99 S1200 M03;	设定恒转速控制指令,进给量单位为mm/r,主轴转速为1200r/min,正转
	G0 X27 Z2 M8;	快速定位到循环起点(X27,Z2)处,打开冷却液
	G71 U1.5 R0.5;	调用纵向粗加工循环指令,背吃刀量为1.5mm,退刀量为0.5mm
	G71 P10 Q20 U0.4 W0.05 F0.2;	径向精加工余量为0.4mm(单边0.2mm),轴向精加工余量为0.05mm,进给量为0.2mm/r
N10	G0 X0;	
	G1 Z0;	
	X9;	
	X10 Z-0.5;	
	Z-15;	
	X19;	
	X20 Z-15.5;	
	Z-30;	
N20	X27;	
	G0 X100 Z150;	快速退刀至(X100,Z150)处
	M5;	停转主轴
	M9;	关闭冷却液
	M01;	选择性暂停(需要按下选择性暂停按钮才起作用),用于观察粗加工完成情况
N2	T0101;	调用外圆车刀
	G97 G99 S1500 M3;	设定恒转速控制指令,进给量单位为mm/r,主轴转速为1500r/min,正转
	G0 X27 Z2 M8;	快速定位到循环起点(X27,Z2)处,打开冷却液
	G70 P10 Q20 F0.1;	调用精加工循环指令,进给量为0.1mm/r
	G0 X100 Z150;	快速退刀至(X100,Z150)处
	M5;	停转主轴
	M9;	关闭冷却液
	M01;	选择性暂停(需要按下选择性暂停按钮才起作用),用于观察粗加工完成情况
N3	T0202;	调用外切槽刀(刀宽3mm)
	G97 G99 S700 M3;	设定恒转速控制指令,进给量单位为mm/r,主轴转速为700r/min,正转
	G0 X22 Z2;	快速定位到(X22,Z2)处

序号	程 序 语 句	注 　 解
	Z-29 M8;	快速定位到 Z-29 处,打开冷却液
	G1 X19 F0.07;	预切槽
	X21 F0.3;	退刀
	Z-27;	
	X19 Z-28 F0.07;	利用刀具右侧刀尖加工 C0.5 倒角
	X0;	切断,根据现场情况也可选择加工至 X2 处,然后手动掰断
	G0 X100;	径向退刀
	Z150;	轴向退刀
	M5;	停转主轴
	M9;	关闭冷却液
	M30;	程序结束

任务三　上机训练

1.10　设备与用具

设备:CK6150e 数控车床。
刀具:外圆车刀、切断刀(刀宽 3mm)。
夹具:三爪自定心卡盘。
工具:卡盘扳手、刀架扳手等。
量具:0~150mm 游标卡尺、0~25mm 外径千分尺。
毛坯:$\phi 25 \times 60$。
辅助用品:垫刀片、毛刷等。

1.11　认识机床

1.11.1　开机

检查机床外观各部位(如防护罩、脚踏板等部位)是否存在异常;检查机床润滑油、冷却液是否充足;检查刀架、夹具、导轨护板上是否有异物;检查机床面板各旋钮状态是否正常;在开机后,检查机床是否存在报警等。可参考表 1-8 对机床状态进行点检。

表 1-8　机床开机准备卡片

检查项目		检查结果	异常描述
机械部分	主轴部分		
	进给部分		
	刀架		
	三爪自定心卡盘		

续表

检 查 项 目		检 查 结 果	异 常 描 述
电器部分	主电源		
	冷却风扇		
数控系统	电气元件		
	控制部分		
	驱动部分		
辅助部分	冷却系统		
	压缩空气		
	润滑系统		

在机床开机前点检正常后,可通过旋转机床后面电气开关 ,打开机床电源。

根据按钮标识方向旋转紧急停止开关 ,解决紧急状态。开机后系统面板如图1-15所示。

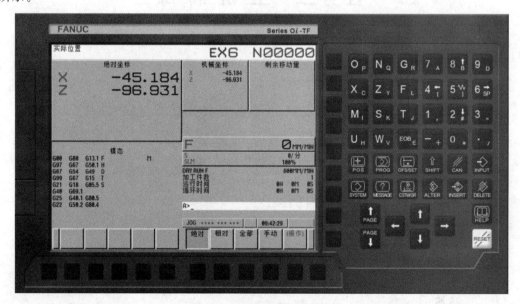

图1-15 启动后的系统面板

按下机床控制面板上的电源开按键 ,使数控系统上电。

1.11.2 机床面板

在正式操作机床前,应熟悉数控车床操作面板各按键的功能及操作方法,熟记紧急按键的具体位置。

1. 操作模式选择按键

图1-16所示为机床的操作模式选择按键,在选定操作模式后,在显示屏上会有相应的标识。

(1)"编辑方式"选择按键:用于编辑程序或外部数控读入。

(2)"自动方式"选择按键：用于自动运行读入内存的程序。

(3)"MDI方式"选择按键：用于运行控制器MDI面板录入的程序。

(4)"手摇方式"选择按键：用于使用电子手轮运动机床轴。

(5)"手动方式"选择按键：使用操作面板方向按键快速移动机床坐标轴。

(6)"回参考点"选择按键：用于机床返回参考点。

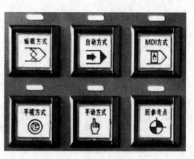

图1-16 操作模式选择按键

2. 其他功能按键

操作面板其他常用功能按键如表1-9所示。由于机床控制系统、生产厂家、出厂批次的不同,因此操作面板功能按键位置、数量会有一些差异。在操作机床前,应熟悉当前机床的操作说明书,并按安全操作要求使用机床。

表1-9 操作面板其他常用功能按键

图 示	说 明
	名称：程序启动功能按键 在自动运转模式(手动输入、记忆、联机)下,选中需要执行的加工程序,按下程序启动功能按键后,程序开始执行
	名称：程序暂停功能按键 (1)在自动运转模式(手动输入、记忆、联机)下,按下程序暂停功能按键后,各轴立即减速停止,进入运转休止状态 (2)当再次按下程序启动功能按键后,加工程序将从当前暂停的程序单段继续执行
	名称：程序保护开关 (1)为防止本机床控制器中的程序被他人编辑、取消、修改、建立,钥匙应交由专人保管 (2)在一般情况下,将此钥匙设定在OFF的位置,以确保程序不被修改或删除 (3)如果想对程序予以编辑、取消或修改,则应将此钥匙设定在ON的位置
	名称："单段"功能按键 本功能按键仅在自动相关模式下有效 (1)当功能按键指示灯亮时,"单段"功能按键有效。此功能打开后,程序将按单段执行,执行完当前单段后程序暂停。在继续按程序启动功能按键后,方可执行下一单段程序,以后执行程序以此类推 (2)当功能按键指示灯不亮时,"单段"功能按键无效。加工程序将一直被执行到程序结束

续表

图　示	说　明
	名称："空运行"功能按键 本功能仅在自动相关模式下有效 （1）当功能按键指示灯亮时，"空运行"功能按键有效。在此功能打开后，程序中所设定的 F 值（切削进给率）指令无效，其各轴移动速率依慢速位移速率指定速率位移 （2）在此功能有效时，若程序执行循环程序，则慢速进给率或切削进给率无法改变，依然按照控制中的 F 值以固定进给率位移
	名称："跳步"功能按键 本功能仅在自动相关模式下有效 （1）当功能按键指示灯亮时，"跳步"功能按键有效。在此功能打开后，在自动运行中，当程序段的开头指定了一个"/"（斜线）符号时，此程序段将略过，不被执行 （2）当功能按键指示灯不亮时，"跳步"功能按键无效。在此功能关闭后，当程序单节前有"/"（斜线）符号时，此程序段也可正常执行
	名称："选择停"功能按键 本功能仅在自动模式有效 （1）当功能按键指示灯亮时，"选择停"功能按键有效。在此功能打开后，在执行程序中，当有 M01 指令时，程序将停止于该单段。若想要继续执行程序，则按下程序启动功能按键即可 （2）当功能按键指示灯不亮时，"选择停"功能按键无效。在此功能关闭后，即使程序中有 M01 指令，程序也不会停止执行
	名称："机床锁住"功能按键 （1）当功能按键指示灯亮时，所有轴机械锁定功能按键有效。在此功能打开后，无论在"手动方式"或"自动方式"中移动任意一个轴，计算机数控（computer numberical control，CNC）均停止向该轴伺服电机输出脉冲（移动指令），但依然在进行指令分配，对应轴的绝对坐标和相对坐标也相应得到更新 （2）M、S、T 代码会继续执行，不受"机床锁住"功能按键限制 （3）在解除此功能后，需要重新回归机械零点，在返回参考点正确且完毕后，再进行其他相关操作 如果未回零点而进行了相关操作，则会造成坐标偏移，甚至出现撞机、程序乱跑等异常现象，从而导致危险
	名称："F1"功能按键 预留功能按键。此功能按键可依据机床实际配置进行设定，工作人员不能自行操作
	名称："F2"功能按键 预留功能按键。此功能按键可依据机床实际配置进行设定，工作人员不能自行操作

续表

图　示	说　明
	名称："F3"（工作灯扩展）功能按键 控制工作灯开启和关闭，不受任何操作模式的限制
	名称："主轴降速"功能按键 (1) 此功能按键位于本机床操作面板上，用于降低编程制定的主轴转速 S，实际转速＝编程给定 S 指令值×主轴速度降低倍率值 (2) 配合主轴控制功能按键使用
	名称："主轴升速"功能按键 (1) 此功能按键位于本机床操作面板上，用于提高编程制定的主轴转速 S，实际转速＝编程给定 S 指令值×主轴速度降低倍率值 (2) 当编程设定速度超过主轴最高转速，且转速达到100％以上倍率时，主轴修调速度等于主轴最高转速 (3) 配合主轴控制功能按键使用
	名称："主轴正转"功能按键 (1) 在本机床上执行一次 S 代码后，选中"手动方式"，按下"主轴正转"功能按键，主轴进行顺时针旋转。主轴旋转速度＝先前执行的主轴速度 S 值×主轴修调旋钮所在的挡位 (2) 使用条件如下。 ① 本按键仅在"手动"模式、"快速"模式、"寸动"模式下，才能使用 ② 在"自动"模式下，当程序中执行主轴正转指令 M03 时，本功能按键指示灯会亮 (3) 当"主轴停止"功能按键或"主轴反转"功能按键生效时，本功能按键指示灯熄灭 (4) 当需要进行主轴反向旋转时，必须先使主轴停止，才可指定反向旋转操作
	名称："主轴停止"功能按键 (1) 主轴无论处于正转或反转状态，按此功能按键均可以停止正在旋转中的主轴 (2) 使用条件如下。 ① 本功能按键仅在"手动"模式、"快速"模式、"寸动"模式下，才能使用 ② 在"自动"模式下，无效 (3) 当主轴停止时，本功能按键指示灯会亮；但在"主轴正转"功能按键或"主轴反转"功能按键生效时，本功能按键指示灯熄灭

续表

图　示	说　明
	名称："主轴反转"功能按键 （1）在本机床上执行一次 S 代码后，选中"手动方式"模式，按下"主轴反转"功能按键，主轴进行逆时针旋转。主轴旋转速度＝先前执行的主轴速度 S 值×主轴修调旋钮所在的挡位 （2）使用条件如下 ① 仅在"手动"模式、"快速"模式、"寸动"模式下，才能使用 ② 在"自动"模式下，当程序中执行主轴反转指令 M04 时，本功能按键指示灯会亮。 （3）当主轴反转时，本功能按键指示灯会亮；但在"主轴正转"功能按键或"主轴停止"功能按键生效时，本功能按键指示灯熄灭 （4）当需要进行主轴正向旋转时，必须先使主轴停止，才可指定正向旋转操作
	名称："冷却"功能按键 （1）在"手动""快速""寸动"模式下，按下此功能按键指示灯亮，冷却液开启 （2）按 RESET 键，冷却液停止喷出，本功能按键指示灯熄灭 （3）在冷却液开启时，需注意冷却液喷嘴的朝向
	名称："手动选刀"功能按键 在"手动""快速""寸动"模式下，每按此功能按键一次，刀具按加方向旋转一个刀位
	名称：＋X 控制功能按键 在 JOG 方式下，按此功能按键，X 轴依进给倍率/快速倍率的速度向机床 X 轴"＋"方向（正方向）移动，同时，本功能按键指示灯点亮；当松开此功能按键后，X 轴停止向机床 X 轴"＋"方向移动，同时，本功能按键指示灯熄灭 此外，本功能按键也作为 X 轴回零点触发键
	名称：－X 控制功能按键 在 JOG 方式下，按此功能按键，X 轴依进给倍率/快速倍率的速度向机床 X 轴"－"方向（负方向）移动，同时，本功能按键指示灯点亮；当松开此功能按键后，X 轴停止向机床 X 轴"－"方向移动，同时，本功能按键指示灯熄灭
	名称：＋Z 控制功能按键 在 JOG 方式下，按此功能按键，Z 轴依进给倍率/快速倍率的速度向机床 Z 轴"＋"方向（正方向）移动，同时，本功能按键指示灯点亮；当松开此功能按键后，Z 轴停止向机床 Z 轴"＋"方向移动，同时，本功能按键指示灯熄灭 此外，本功能按键也作为 Z 轴回零点触发键
	名称：－Z 控制功能按键 在 JOG 方式下，按此功能按键，Z 轴依进给倍率/快速倍率的速度向机床 Z 轴"－"方向（负方向）移动，同时，本功能按键指示灯点亮；当松开此功能按键后，Z 轴停止向机床 Z 轴"－"方向移动，同时，本功能按键指示灯熄灭 此外，当程序执行 Z 轴负方向移动程序指令时，该功能按键指示灯也将点亮；停止该移动指令时，该功能按键指示灯熄灭

续表

图　　示	说　　明
	名称："超程释放"功能按键 (1) 当本机床各轴的行程超过硬体极限时,机床会出现超程报警,机床的动作将停止。这时,按住这个功能按键,在"手摇方式"下用手持单元将机床超程的轴反方向移动 (2) 绝对式编码器机床超程无须按此功能按键
	名称：手动快移功能按键 本功能仅在"手动"模式下有效。在"手动"模式下,按下此功能按键,指示灯点亮。在"手动"模式下,实际快速进给速度＝参数设置 G00 指令最大速度值×快速倍率开关所在的倍率值％
	名称：进给倍率及进给修调旋钮 (1) 此旋钮位于本机床操作面板上,控制编程指定 G01 指令速度,实际进给速度＝编程给定 F 指令值×进给倍率开关所在倍率值％ (2) 在"寸动"模式下,此时控制 JOG 进给倍率,实际 JOG 进给速度＝参数设定固定值×进给倍率开关所在倍率值％ (3) 配合轴进给控制功能按键使用
	名称：快速倍率功能按键 (1) 此功能按键位于本机床操作面板上,控制编程指定 G00 指令速度,实际进给速度＝参数设置 G00 指令最大速度值×快速倍率功能按键所在倍率值％ (2) 在"快送"模式下,此时控制手动快速进给倍率,实际进给速度＝参数设置 G00 指令最大速度值×快速进给倍率开关所在倍率值％。快速倍率可以在 F0、25％、50％、100％ 4 个挡位调整 (3) 配合轴进给控制功能按键使用

1.11.3　回参考点

数控车床各进给轴如果采用增量编码器,则在系统启动后,需要手动使各轴回参考点。先按下"回参考点"选择按键，然后按操作面板＋X 控制功能按键，再按＋Z 控制功能按键，使各轴返回到参考点,以确定机床坐标原点。在执行返回参考点操作时,应先返回 X 轴,再返回 Z 轴,以免发生碰撞。机床如果采用绝对编码器,则在开机后不需要执行返回参考点的操作。

1.12　刀具准备

在加工前,应先将本任务所需刀具准备齐全,所需要的刀具如表 1-10 所示,按其中序列,正确安装刀具。

表 1-10 刀具安装

刀 号	刀 柄	刀 片	安装工具
T1			
T2			

在安装刀具时,应使用正确的工具及方法安装,错误操作可能损坏刀具、刀柄,甚至造成人员伤害。刀具的安装精度对加工精度及刀具的使用寿命也有较大的影响。

1.13 设定工件原点

工件原点设定的过程也称对刀,正确的原点设置对于数控加工来说至关重要。对刀按照自动化程度可分为自动对刀和手动对刀。自动对刀需要设备具有自动对刀装置,应用较少;手动对刀通常使用试切法,应用较为广泛。

1.13.1 Z 轴原点设定

首先,根据加工要求安装好毛坯,保证牢固可靠。然后,将机床操作模式切换到"MDI 方式"手动数据输入状态,在系统面板 MDI 程序输入位置输入程序段"T0101;",按循环启动键运行当前程序段,机床会执行换刀操作。

再输入程序段"S500 M3;",按循环启动键运行当前程序段,使主轴以 500r/min 的速度低速旋转。

在"手动方式"下,点动 X 轴、Z 轴方向控制键,使刀具移动到工件的左侧附近。然后切换到"手摇方式"操作模式,用手轮缓慢移动。在准备切削之前,需要将手轮倍率调整为"×10",如果由于倍率过大致使刀具运动速度过快,则会导致刀具和工件发生损坏。控制刀具沿 X 轴方向(径向)微量切削毛坯端面,在切削完成后,再沿 X 轴方向退回,注意此时不能移动 Z 轴。

此时,按系统面板上 [OFS/SET] 按键,单击显示屏底部 [刀偏] 按钮,再单击 [形状] 按钮,移动光标至对应刀号位置,输入"Z0;",单击显示屏底部的"测量"按钮,完成该刀具的 Z 轴方向对刀。

1.13.2 X 轴原点设定

利用外圆车刀微量车削工件外圆表面,车削长度能够满足测量即可,然后将刀具沿着 Z 轴(轴向)退刀,注意此时不能移动 X 轴。停止主轴转动,利用游标卡尺或千分尺测量已加工的外圆直径,如测得外圆直径为 24.36mm。

此时,按系统面板上 按键,单击显示屏底部 [刀偏] 按钮,再单击 [形状] 按钮,移动光标至对应刀号位置,输入"X24.36;",单击显示屏底部的"测量"按钮,完成该刀具的 X 方向对刀。系统界面如图 1-17 所示。通常情况下,工件直径的精度要求较高,所以还需要在对应的"磨损"位置输入合适的补偿值,以便后期通过修改该磨损补偿值控制尺寸精度。

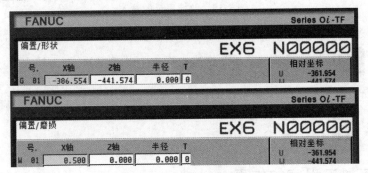

图 1-17　刀具补偿界面

测量其他刀具的方法类似。但需要注意的是,应首先将刀架移动到安全位置,然后在"MDI 方式"下换刀。在 Z 轴方向对刀时,不能使用其他刀具再次车削零件端面,只需控制刀具与已经加工完成的端面刚好接触即可,以保证各不同刀具原点位置的一致性。切断刀由于结构限制,导致该刀具沿 Z 轴切削时的强度较差,所以,切断刀沿轴向不能进行大切深的切削,只能进行微量加工,以免刀具发生损坏。

1.14　程序编辑

在机床系统面板上进行程序的录入共有两种操作,一种是程序号的录入,另一种是程序语句的录入。

在操作面板上将操作模式切换到"编辑"模式,在系统面板上按下 [PROG] 按键,在显示屏底部 [程序 目录 下一步 程序检 操作] 可通过单击"目录"按钮切换程序存放的路径,通过单击"程序"按钮进行程序的编辑。

在录入程序时,首先单击"程序"按钮进入程序的编辑界面。

先输入程序号 [Oₚ] 程序号(范围为 0001~8999)后,直接按下 [INSERT] 按键,再按下 [EOB_E] 按键和 [INSERT] 按键实现换行。此时要注意,程序号(名)和程序段结束符 [EOB_E] 是分两步输入的,如果输入"O0001;",则按下 [INSERT] 按键会提示"格式错误" A)O0001;_ 格式错误。

程序体其他语句的输入与程序名不同,在直接输入整段语句后,直接按下 [INSERT] 按键即可,如 [N₀1,T₀1₀1,EOB_E INSERT]。

完成输入的程序如图 1-18 所示。

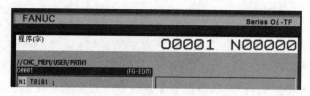

图 1-18 数控程序的输入

如在编辑过程中需要删除程序块,则将光标移动到要删除的程序块位置(黄色高亮显示),然后按下 ![DELETE] 按键完成删除;如需修改光标选中的程序块,则需输入欲更改后的内容,再按下 ![ALTER] 按键即可完成程序块内容的替换;如需删除待输入区 ![输入区] 中的最后一个字符,则按下 ![CAN] 按键即可实现退格。

1.15 数控加工程序的仿真

FANUC-0i-MF 系统提供了数控加工程序的图形校验功能,由于系统版本和机床生产商出厂配置的不同,因此,图形校验界面也有差异。

在操作面板上,选择 AUTO 模式,即程序的自动执行模式,调入需要进行图形校验的数控加工程序,然后在系统面板上按下 ![CSTMGR] 按键,进入图形校验界面,如图 1-19 所示,可通过图形校验界面参数设备仿真区域大小和视角。

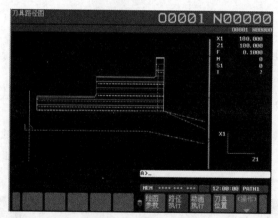

图 1-19 图形校验界面

在操作面板上按下 ![MC LOCK] 按键和 ![DRY RUN] 按键,开启程序的坐标轴锁定状态和空运行状态,按循环启动功能按键,程序即开始图形校验,通过调节进给倍率旋钮可调整图形校验速度。

1.16 零件加工

在图形校验过程验证无问题后,即可进行零件加工。在零件加工前,应详细了解机床的安全操作要求,穿戴好劳动保护服装和用具。在进行零件加工时,应熟悉数控车床各操作按键的功能和位置，了解紧急状况的处置方法。

由于此次操作属于首件加工，所以存在一定的风险，操作人员需要全程保持注意力高度集中。在自动加工前，需要将快速倍率旋钮调至最小，进给倍率调至"0"，防止在启动后，失去对机床的有效控制。检查循环起点位置的正确性尤为重要，若发现刀具位置与坐标不符，则需立即停止运行，以防发生危险。在自动运行过程中，应认真观察加工状况，如表面质量和切屑排除情况等，若发生异常，则应及时进行人工干预。

注意：如已进行图形校验操作，则在操作完成后，须执行机床回归机械零点操作，然后再进行其他的相关操作。如果未回零点而进行了相关操作，则会造成坐标偏移，甚至出现撞机、程序乱跑等异常现象，从而导致危险。

在操作面板上，选择 AUTO 模式，即程序的自动执行模式，调入需要进行加工的数控加工程序，按下循环启动按键进行自动加工，如图 1-20 所示。

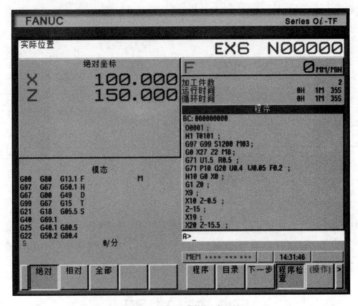

图 1-20　程序的自动运行

在此状态下，每按下一次循环启动按键，程序都只自动执行光标所在的一行。在按下循环启动功能按键前，应观察刀具与工件间的距离是否安全；在按下循环启动后，通过进给倍率旋钮控制机床的运动速度，同时，对照显示屏"余移动量"栏显示的剩余移动量，观察刀具与工件之间的实际距离。若实际距离与剩余移动量相差过大，则应果断停机检查，以免发生撞机事故。在程序调试过程中，还应密切注意显示屏的"模态"状态显示，确保主轴转速、进给速度、工件坐标系号、补偿状态及补偿号等无异常发生。

1.17　零件检测

在零件加工完成后，应当认真清理工件，并按照质量管理的相关要求，对加工完成的零件进行相关检验，保证生产质量。机械加工零件"三级"检验卡片，如表 1-11 所示。

表1-11 机械加工零件"三级"检验卡片

零部件图号		零部件名称		工 序 号	
材料		送检日期		工序名称	
检验项目	自检结果	互检结果	专业检验	备注	
检验结论	□合格　　□不合格　　□返修　　□让步接收 　　　　　　　　　　　　检验签章： 　　　　　　　　　　　　　　　　年　　月　　日				
不符合项描述					

项目总结

阶梯轴作为数控车床的典型加工零件,广泛应用在各种设备中。根据设备情况和精度的要求,其加工工艺也存在一些差别。编程人员及操作人员需要结合加工条件,合理制定加工工艺,以提高零件的加工精度和生产效率。

课后习题

1. 填空题

(1) 数控机床主要由_____和_____两部分组成。

(2) 数控机床按其控制刀具与工件相对运动的方式分为_____数控机床、_____数控机床、_____数控机床。

(3) 数控车床的加工动作主要分为_____的运动和_____的运动两部分。

(4) 数控机床上的坐标系采用_____坐标系。

(5) 数控编程一般分为_____和_____。

(6) 数控车床关机一般先按下_____后,再按下_____,最后关闭机床总电源。

(7) 数控车床工件原点一般设在_____与右端面(或左端面)的交点处,X轴的坐标值取_____尺寸。

2．判断题

（1）刀具在某一坐标轴方向上远离工件的方向为该坐标轴的负方向。（   ）

（2）G98 为每转进给控制指令，G99 为每分钟进给控制指令。（   ）

（3）在数控车床程序中，运用主轴恒线速控制指令 G96 可以提高工件表面质量的一致性。（   ）

（4）车削指令 G50 除了具有坐标系设定功能以外，还可以用于限定主轴的最高转速。（   ）

（5）在数控车床编程时，采用直径尺寸编程。（   ）

（6）在按下机床操作控制面板上的"选择停止"按键后，M01 指令的执行过程和 M00 指令的执行过程相同。（   ）

（7）FANUC 系统中 00 组的 G 指令都是非模态指令。（   ）

3．选择题

（1）在数控机床中，（   ）是由传递切削动力的主轴所决定。

    A．X 轴坐标   B．Z 轴坐标   C．Y 轴坐标   D．C 轴坐标

（2）数控零件加工程序输入必须在（   ）工作方式下进行。

    A．手动   B．自动   C．手动数据输入   D．编辑

（3）车床上的三爪自定心卡盘和铣床上的平口钳属于（   ）。

    A．通用夹具   B．专用夹具   C．组合夹具   D．随行夹具

（4）G00 指令的移动速度值由（   ）指定。

    A．机床参数   B．数控程序   C．操作面板   D．随机

（5）在下列轨迹中，G00 的轨迹可能是（   ）。

    A．直线   B．斜直线   C．折线   D．以上皆有可能

（6）当确定数控机床坐标轴时，一般应先确定（   ）。

    A．X 轴   B．Y 轴   C．Z 轴   D．A 轴

（7）功能按键"F0""F25""F50""F100"用于控制数控机床的（   ）倍率。

    A．手动进给   B．自动进给   C．增量进给   D．快速进给

（8）在执行完程序段"G00 X20.0 Z30.0；G01 X10.0 W20.0 F0.2；U-40.0 W-70.0；"后，刀具所达到的工件坐标系的位置为（   ）。

    A．（X−40.0, Z−70.0）   B．（X−30.0, Z−50.0）

    C．（X−30.0, Z−20.0）   D．（X−10.0, Z−20.0）

4．简答题

（1）简述数控车床夹的种类。

（2）简述刀具刀片材料的选用原则。

（3）简述纵向粗加工循环指令 G71 在车削外圆时，需要注意什么。

5．综合编程题

根据图 1-21 所示的零件图确定加工工艺、编写程序，并自动加工。

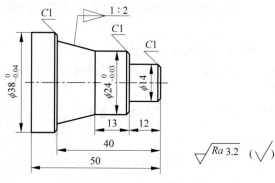

图 1-21 题 5 图

自我学习检测评分表如表 1-12 所示。

表 1-12 自我学习检测评分表

项　　目	目标要求	分值	评分细则	得分	备注
学习关键知识点	（1）了解数控车床的分类，以及 CK6150e 数控车床的结构及主要参数 （2）理解车削加工的特点 （3）熟悉常用刀具的分类，并能进行刀具的正确选择 （4）了解数控车床的通用夹具及量具 （5）掌握数控车削程序基本格式及基本准备功能指令	20	理解与掌握		
工艺准备	（1）能够正确识读轴类零件图 （2）能够独立确定加工工艺过程，并正确填写工艺文件 （3）能够根据工序加工工艺，编写正确的加工程序	30	理解与掌握		
上机训练	（1）会正确选择相应的设备与用具 （2）掌握外圆车刀和切断刀的选择和安装方法 （3）能够正确操作数控车床，并根据加工情况调整加工参数	50	（1）理解与掌握 （2）操作流程		

思政小课堂

项目二　手柄车削编程与加工训练

▶ 思维导图

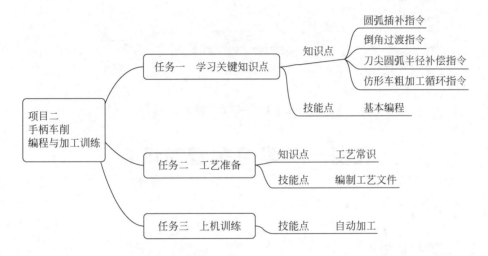

▶ 学习目标

知识目标
(1) 了解曲线轮廓的加工特点。
(2) 理解仿形车粗加工循环指令各参数的含义。

能力目标
(1) 能够独立确定加工工艺路线,并正确填写工艺文件。
(2) 能够正确操作数控车床,并根据加工情况调整加工参数。
(3) 能够根据零件结构特点和精度合理选用量具,并正确、规范地测量相关尺寸。

素养目标
(1) 培养学生的科学探究精神和态度。
(2) 培养学生的工程意识。
(3) 培养学生的团队合作能力。

手柄零件主要用于工作人员手持操作,通常由连接部分和手持部分组成。连接部分用于与其他零件进行连接,并传递运动;手持部分用于工作人员手动操作。手持部分由于在工作过程中与手直接接触,所以通常要求表面光整、无毛刺,以免造成不必要的人身伤害。手柄零件作为手动操作零件在生产及生活中应用广泛。

▶任务引入

根据手柄零件图(见图 2-1)要求制定加工工艺,编写数控加工程序,并完成手柄零件的加工。该零件毛坯材料为 45 钢,调质处理,要求表面光整。

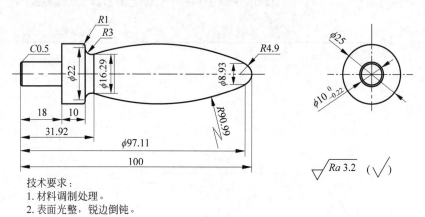

技术要求:
1. 材料调制处理。
2. 表面光整,锐边倒钝。

图 2-1 手柄零件图

任务一 学习关键知识点

2.1 基本指令

2.1.1 圆弧插补指令

圆弧插补指令是将刀具以圆弧的形式按 F 代码指定的速率,从它的当前位置移动到指令要求的位置,其特点是各坐标以联动的方式,且按进给速度 F 作任意圆弧运动。圆弧插补根据圆弧运动方向又分为顺时针圆弧插补和逆时针圆弧插补,指令分别为 G02 指令和 G03 指令。在编程时,用 X、Z(或 U、W)指定圆弧终点坐标。用 R __ 指定圆弧半径(当圆心角小于或等于 180°时,R 为正值;当圆心角大于 180°,且圆心角小于 360°时,R 为负值),或者用 I __ K __ 指定圆弧圆心相对于圆弧起点的坐标增量,当 **I**、**K** 为 0 时,可以省略。

指令格式为:

G02 X __ Z __ R __ F __ ;(顺时针圆弧插补)
G03 X __ Z __ R __ F __ ;(逆时针圆弧插补)

或

G02 X __ Z __ I __ K __ F __ ;
G02 X __ Z __ I __ K __ F __ ;

如图 2-2 所示,刀具由点 A 以指定的速度 F 逆时针移动到点 B,再由点 B 以指定的速度 F 顺时针移动到点 C。

利用圆弧插补指令编程如表 2-1 所示。

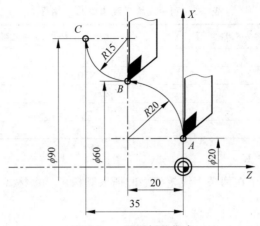

图 2-2　圆弧插补指令

表 2-1　利用圆弧插补指令编程

利用半径 R 编程	利用圆心矢量 I、K 编程
…	…
G03 X60 Z-20 R20 F0.2;	G03 X60 Z-20 I0 K-20 F0.2;
G02 X90 Z-35 R15;	G02 X90 Z-35 I15 K0;
…	…

2.1.2　倒角过渡指令

在车削零件时,锐边通常需要进行倒棱或倒圆角处理,以利于后期装配及安全要求。利用倒角过渡指令编程可以简化程序的编写。

1. 倒棱过渡指令 C＿

倒棱过渡指令用于在两条直线轮廓之间插入倒棱,指令格式为:

G01 X＿ Z＿ C＿ F＿;

例如,在编制如图 2-3 所示外圆轮廓时,可以直接对交点 D 进行编程,无须编制直线 AB 轮廓段。

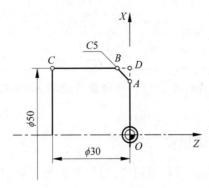

图 2-3　倒棱过渡指令

利用倒棱过渡指令 C＿编程如表 2-2 所示。

表 2-2　利用倒棱过渡指令 C__ 编程

利用直线插补指令 G01 编程	利用倒棱过渡指令 C__ 编程
…	…
G01 X40；	G01 X50 C5；
X50 Z-5；	Z-30；
Z-30；	…
…	…

2. 倒圆角过渡指令 R__

倒圆角过渡指令用于在两条直线轮廓之间插入圆角。指令格式为：

G01 X__ Z__ R__ F__；

例如，在编制如图 2-4 所示外圆轮廓时，可以直接对交点 D 进行编程，无须编制圆弧 AB 轮廓段。

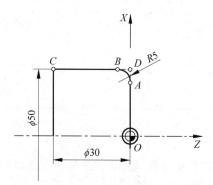

图 2-4　倒圆角过渡指令

利用倒圆角过渡指令 R__ 编程如表 2-3 所示。

表 2-3　利用倒圆角过渡指令 R__ 编程

利用直线插补指令 G01 编程	利用倒圆角过渡指令 R__ 编程
…	…
G01 X40 F0.1；	G01 X50 R5 F0.1；
X50 Z-5；	Z-30；
Z-30；	…
…	

注意：倒圆角过渡编程只适用于相交的两条直线之间，且直线实际长度不能为"0"，编程终点坐标为交点坐标。

2.1.3　刀尖圆弧半径补偿指令

在数控车削加工时，为提高刀具耐用度，通常刀尖都会制作成圆弧形。如图 2-5 所示，如果在没有建立刀尖圆弧半径补偿的情况下，运用带有圆弧刀尖的刀具加工除了端面和圆柱表面以外的其他轮廓，则会造成过切或欠切现象。为了解决这个问题，各种数控系统都引入了刀尖圆弧半径补偿指令。

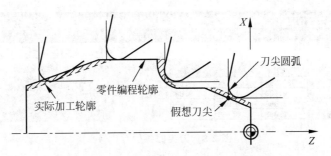

图 2-5　刀尖圆弧半径补偿对加工的影响

不同刀具的圆弧位置不同，需要补偿的方向也不同，为了区别各种不同刀具，引入了刀尖圆弧半径位置代号，如图 2-6 所示。在建立和取消刀尖圆弧半径补偿时，需要与直线插补指令 G01 或快速定位指令 G00 配合使用。补偿的实施分为三个过程：建立补偿是在刀具进入零件轮廓之前完成的；执行补偿是刀具在零件轮廓上连续加工；补偿取消是在刀具离开轮廓之后实施的。

（1）刀尖圆弧半径左补偿指令 G41（见图 2-7），指令格式为：

G01(G00)G41 X__ Z__ F__；

（2）刀尖圆弧半径右补偿指令 G42（见图 2-7），指令格式为：

G01(G00)G42 X__ Z__ F__；

（3）取消刀尖圆弧半径补偿指令 G40，指令格式为：

G01(G00)G40 X__ Z__ F__；

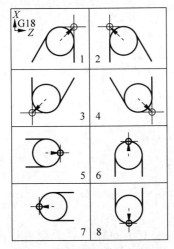

图 2-6　刀尖圆弧半径位置代号

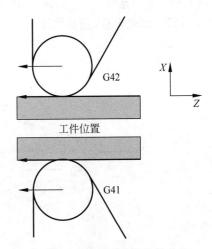

图 2-7　刀尖圆弧半径补偿方向

在机床中，还需要在刀具偏置参数中输入对应的刀尖圆弧位置代号，如图 2-8 所示。例如，外圆车刀的代号为 3，将其输入对应的位置中。

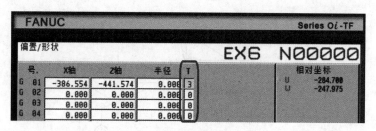

图 2-8　刀尖圆弧位置代号

2.1.4　仿形粗加工循环指令 G73

指令格式为：

G73 U (Δi) W (Δk) R (d);
G73 P (ns) Q (nf) U (Δu) W (Δw) F(f);

其中，Δi 表示刀具沿着 X 轴（径向）的退刀量，即在该方向的总待加工余量；Δk 表示刀具沿着 Z 轴（径向）的退刀量，即在该方向的总待加工余量；Δd 表示粗加工次数；ns 表示精加工轮廓程序段中的开始程序段号；nf 表示精加工轮廓程序段中的结束程序段号；Δu 表示 X 方向精加工余量，加工内轮廓时为负值；Δw 表示 Z 方向精加工余量，一般取 0.05～0.1mm。

G73 指令主要用于铸件、锻件和径向尺寸非单调变化的轮廓切削加工。对于毛坯与零件轮廓极为接近的情况，加工效率较高；但是，对于圆柱形毛坯进行多刀加工的情况，空走刀路径较多，效率也就相对较低。G73 指令动作如图 2-9 所示，刀具首先从循环起点沿着 X 轴和 Z 轴负向同时快速移动 $\Delta i + \Delta u/2$ 和 $\Delta k + \Delta w$ 的距离，进刀，按照轮廓轨迹切削加工，退刀，快速返回。以此类推，完成粗加工。

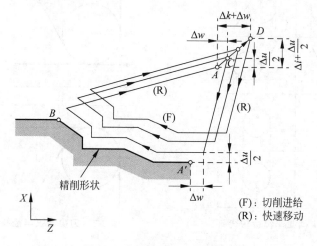

图 2-9　仿形车粗加工循环指令

在使用 G73 指令粗加工时，包含在 ns～nf 程序段中的 F、S 代码对粗车循环无效，在顺序号为 ns～nf 的程序段中，不能调用子程序。

任务二 工艺准备

2.2 零件图分析

根据零件的使用要求,选择45钢作为手柄零件的毛坯材料,毛坯下料尺寸定为$\phi 30 \times 105$。在加工时,以$\phi 30$毛坯外圆作为粗基准,粗、精加工右侧手柄部分至要求尺寸,然后掉头装夹在$\phi 25$外圆处(在装夹时注意做好保护,以防表面夹伤),加工零件左端$\phi 10$和$\phi 25$阶梯轴端面及外圆至要求尺寸。

注意:在车削$\phi 25$外圆时,车削长度要足够。另外,在装夹毛坯时,应注意棒料伸出的长度,以免刀具与卡盘发生碰撞。

2.3 工艺设计

根据零件图分析,确定工艺过程,如表2-4所示。

表2-4 工艺过程卡片

机械加工工艺过程卡片		产品型号	CLJG-01	零部件序号	HS-01	第1页	
		产品名称	槽轮机构	零部件名称	手柄	共1页	
材料牌号	C45	毛坯规格	$\phi 30 \times 105$	毛坯质量	kg	数量	1
工序号	工序名	工序内容		工段	工艺装备	工时/min	
						准结	单件
5	备料	按$\phi 30 \times 105$尺寸备料		外购	锯床		
10	车加工	以$\phi 30$毛坯外圆作为粗基准,粗、精加工手柄部分		车	车床外径千分尺	45	30
15	车加工	以已加工完成的手柄部分及$\phi 25$外圆作为精基准(尽量同轴),粗、精加工$\phi 10$和$\phi 25$外圆表面及端面,并保证长度要求		车	车床外径千分尺	45	30
20	清理	清理工件,锐角倒钝		钳			
25	检验	检验工件尺寸		检			

本训练任务针对第10、第15工序车削加工,进行工序设计,制订工序卡片,如表2-5所示。

表 2-5 车削加工工序卡片

机械加工工序卡片	产品型号	CLJG-01	零部件序号	HS-01	第 1 页
	产品名称	槽轮机构	零部件名称	手柄	共 1 页
			工序号		10、15
			工序名		车加工
			材料		C45
			设备		数控车床
			设备型号		CK6150e
			夹具		三爪卡盘
			量具		游标卡尺
					千分尺
			准结工时		90min
			单件工时		60min

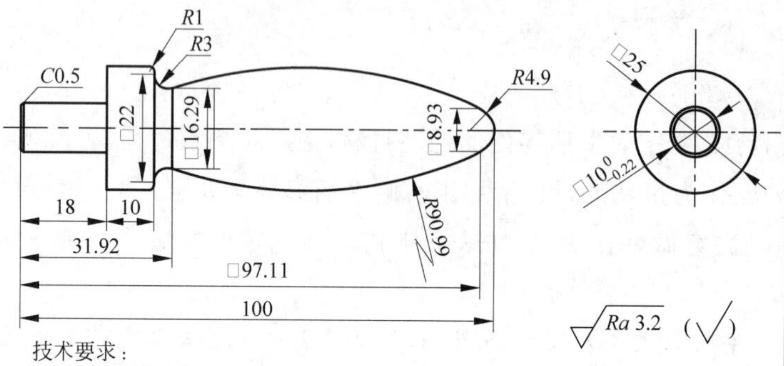

技术要求：
1. 材料调制处理。
2. 表面光整，锐边倒钝。

工步	工步内容	刀具	S/(r/min)	F/(mm/r)	a_p/mm	工步工时/min 机动	工步工时/min 辅助
1	工件安装						5
2	粗加工 ϕ10 和 ϕ20 外圆表面及端面，精加工余量为 0.2mm	外圆粗车刀	1200	0.2	1.5	15	
3	精加工 ϕ10 和 ϕ20 外圆表面及端面	外圆精车刀	1500	0.1	0.2	10	
4	拆卸、清理工件						5

2.4 数控加工程序编写

根据工序加工工艺，编写加工程序，如表 2-6 所示。

表 2-6 手柄数控加工程序

序号	程序语句	注　解
	O0001；	右侧数控加工程序
N1	T0101；	调用外圆车刀
	G97 G99 S1100 M03；	设定恒转速控制，进给量单位为 mm/r，主轴转速为 1000r/min，正转
	G0 X32 Z2 M8；	快速定位到循环起点（X32,Z2）处，打开冷却液
	G73 U15 W0 R10；	调用仿形粗加工循环指令
	G73 P10 Q20 U0.4 W0 F0.16；	径向精加工余量为 0.4mm（单边 0.2mm），轴向精加工余量为 0.05mm，进给量为 0.16mm/r
N10	G0 X0；	
	G1 G42 Z0；	
	G3 X8.93 Z-2.89 R4.9；	

续表

序号	程序语句	注　解
	G3 X16.29 Z-68.08 R90.99;	
	G2 X22 Z-72 R3;	
	G1 X25 R1;	
	Z-83;	
N20	G1 G40 X32;	
	G0 X100 Z150;	快速退刀至(X100,Z150)处
	M5;	停转主轴
	M9;	关闭冷却液
	M01;	选择性暂停(需要按下选择性暂停按钮才起作用),用于观察粗加工完成情况
N2	T00101;	调用外圆车刀
	G97 G99 S1200 M3;	设定恒转速控制,进给量单位为mm/r,主轴转速1200r/min,正转
	G0 X32 Z2 M8;	快速定位到循环起点(X32,Z2)处,打开冷却液
	G70 P10 Q20 F0.1;	调用精加工循环指令,进给量为0.1mm/r
	G0 X100 Z150;	快速退刀至(X100,Z150)处
	M5;	停转主轴
	M9;	关闭冷却液
	M30;	程序结束
	O0002;	**左侧数控加工程序**
N3	T0101;	调用外圆车刀
	G97 G99 S1000 M3;	设定恒转速控制,进给量单位为mm/r,主轴转速为1000r/min,正转
	G0 X32 Z10;	快速定位到(X32,Z10)处
	G71 U1.5 R0.5;	调用纵向粗加工循环指令,背吃刀量为1.5mm,退刀量为0.5mm
	G71 P30 Q40 U0.4 W0.05 F0.18;	径向精加工余量为0.4mm(单边0.2mm),轴向精加工余量为0.05mm,进给量为0.18mm/r
N30	G0 X0;	
	G1 Z0;	
	X10 C0.5;	
	Z-18;	
N40	X27;	
	G0 X100;	径向退刀
	Z150;	轴向退刀
	M5;	停转主轴
	M9;	关闭冷却液
	M01;	选择性暂停(需要按下选择性暂停按钮才起作用),用于观察粗加工完成情况
N4	T00101;	调用外圆车刀
	G97 G99 S1200 M3;	设定恒转速控制,进给量单位为mm/r,主轴转速为1200r/min,正转

续表

序号	程序语句	注　解
	G0 X32 Z2 M8;	快速定位到循环起点(X32,Z2)处,打开冷却液
	G70 P30 Q40 F0.1;	调用精加工循环指令,进给量为0.1mm/r
	G0 X100 Z150;	快速退刀至(X100,Z150)处
	M5;	停转主轴
	M9;	关闭冷却液
	M30;	程序结束

任务三　上机训练

2.5　设备与用具

设备：CK6150e 数控车床。

刀具：外圆车刀、切断刀(刀宽 3mm)。

夹具：三爪自定心卡盘。

工具：卡盘扳手、刀架扳手等。

量具：0～150mm 游标卡尺、0～25mm 外径千分尺。

毛坯：$\phi 30 \times 105$。

辅助用品：垫刀片、毛刷等。

2.6　开机前检查

参考表 2-7 对机床状态进行点检。

表 2-7　机床开机准备卡片

检查项目		检查结果	异常描述
机械部分	主轴部分		
	进给部分		
	刀架		
	三爪自定心卡盘		
电器部分	主电源		
	冷却风扇		
数控系统	电气元件		
	控制部分		
	驱动部分		
辅助部分	冷却系统		
	压缩空气		
	润滑系统		

2.7 加工前准备

在加工前,应先将本任务所需刀具准备齐全,并安装正确。根据工艺要求设定工件原点,录入数控加工程序,并进行图形校验。

2.8 零件加工

在图形校验过程验证无问题后,即可进行零件加工。在零件加工前,应详细了解机床的安全操作要求,穿戴好劳动保护服装和用具。在进行零件加工时,应熟悉数控车床各操作按键的功能和位置,了解紧急状况的处置方法。在加工过程中,尤其是在即将切削之前,应对照显示屏"余移动量"栏显示的剩余移动量,观察刀具与工件之间的实际距离。若实际距离与剩余移动量相差过大,则应果断停机检查,以免发生撞机事故。若有异常,则应及时停止机床运动。

2.9 零件检测

在零件加工完成后,应当认真清理工件,并按照质量管理的相关要求,对加工完成的零件进行相关检验,保证生产质量。机械加工零件"三检"检验卡片如表 2-8 所示。

表 2-8 机械加工零件"三级"检验卡片

零部件图号		零部件名称		工 序 号	
材料		送检日期		工序名称	
检验项目	自检结果	互检结果	专业检验	备注	
检验结论	□合格　　□不合格　　□返修　　□让步接收　　　　　　　　　　检验签章:　　　　　　　　　　　　　　　　年　　月　　日				
不符合项描述					

项目总结

手柄作为数控车床的典型加工零件,在生产和生活中应用广泛。根据设备情况和精度要求,其加工工艺也会存在一些差别。编程人员及操作人员需要结合加工条件,合理制定加工工艺,以提高零件的加工精度和生产效率。

课后习题

1. 填空题

(1) 刀尖圆弧半径补偿的过程分为_____、_____和_____。

(2) 数控车床的刀具补偿有_____和_____两种。

(3) 外圆表面主要加工方法是_____、_____。

(4) 取消刀尖圆弧半径补偿的指令是_____。

(5) 仿形粗加工循环指令 G73 的加工轮廓写在_____之间。

2. 判断题

(1) 圆弧插补用 R 代码编程时,当圆弧的圆心角小于 180°时,R 取负值。(   )

(2) 在 FANUC 系统中,刀尖圆弧半径补偿模式的建立与取消程序段只能在 G00 指令和 G01 指令下才有效。(   )

(3) 若在 FANUC 系统的 G71 指令中 ns～nf 程序段编写了非单调变化的轮廓,则在 G71 指令执行过程中会产生报警。(   )

(4) 圆弧插补指令是指从 Y 轴负方向看,顺时针圆弧插补指令用 G02 表示。(   )

(5) FANUC 系统的倒角过渡指令可用于任何角度的两相交直线的倒角。(   )

(6) 在 G70 指令循环结束后,刀具返回参考点。(   )

3. 选择题

(1) 圆弧插补指令中的 I、K 值是指(   )的矢量值。

    A. 起点到圆心   B. 终点到圆心   C. 圆心到起点   D. 圆心到终点

(2) 在精加工时,选择切削用量一般是以(   )为主。

    A. 提高生产率   B. 降低切削功率   C. 保证加工质量   D. 提高表面质量

(3) 在使用内、外圆复合固定循环进行编程时,在其 ns～nf 之间的程序段中,当含有(   )指令时,程序执行过程中不会产生程序报警。

    A. 固定循环   B. 参考点返回

    C. 复合固定循环   D. 圆心角大于 90°的圆弧加工

(4) 在下列固定循环中,顺序号为 ns 的程序段必须沿 Z 轴方向进刀,且不能出现 X 坐标的固定循环是(   )。

    A. G71   B. G72   C. G73   D. G74

(5) 铸造成型、粗加工成型的工件选用(   )指令作为粗加工循环指令较为合适。

    A. G71   B. G72   C. G73   D. G74

(6) 当使用 G02/G03 指令时,下面关于使用倒圆角过渡指令编程的说明不正确的是( )。

  A. 整圆加工不能采用该方式编程　　B. 该方式与使用 I、J、K 效果相同

  C. 大于 180° 的弧 R 取正值　　D. R 可取正值、负值,但加工轨迹不同

4. 简答题

(1) 简述如何判断刀尖圆弧半径补偿方向。

(2) 简述刀尖圆弧半径补偿有哪些作用。

(3) 简述仿形粗加工循环指令 G73 主要用于哪些场合。

自我学习检测评分表如表 2-9 所示。

表 2-9　自我学习检测评分表

项　目	目 标 要 求	分　值	评分细则	得分	备注
学习关键知识点	(1) 掌握倒角过渡指令的使用 (2) 掌握刀尖圆弧半径补偿指令的使用 (3) 理解仿形粗加工循环指令各参数的含义	20	理解与掌握		
工艺准备	(1) 能够正确识读零件图 (2) 能够独立确定加工工艺路线,并正确填写工艺文件 (3) 能够根据工序加工工艺,编写正确的加工程序	30	理解与掌握		
上机训练	(1) 能够根据零件结构特点和精度合理选用量具,并正确、规范地测量出相关尺寸 (2) 掌握手柄车削加工的操作流程 (3) 能够正确操作数控车床,并根据加工情况调整加工参数	50	(1) 理解与掌握 (2) 操作流程		

思政小课堂

项目三　驱动轴车削编程与加工训练

➢思维导图

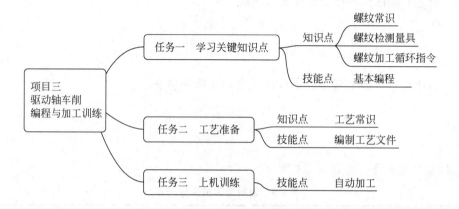

➢学习目标

知识目标

(1) 了解螺纹轴的加工特点。
(2) 理解螺纹加工循环指令中各参数的含义。

能力目标

(1) 能够独立确定加工工艺路线,并正确填写工艺文件。
(2) 能够正确操作数控车床,并根据加工情况调整加工参数。
(3) 能够根据零件结构特点和精度合理选用量具,并正确、规范地测量相关尺寸。

素养目标

(1) 培养学生的科学探究精神和态度。
(2) 培养学生的工程意识。
(3) 培养学生的团队合作能力。

螺纹轴零件广泛地应用于生产和生活中,主要由阶梯轴部分和螺纹部分构成。外螺纹部分用于与螺母连接,起到固定的作用。

➢任务引入

根据驱动轴零件图(见图 3-1)要求制定加工工艺,编写数控加工程序,并完成驱动轴零件的加工。该零件毛坯材料为 45 钢,调质处理,要求表面光整。

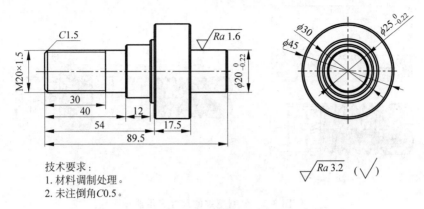

技术要求：
1. 材料调制处理。
2. 未注倒角C0.5。

图 3-1 驱动轴零件图

任务一 学习关键知识点

3.1 螺纹常识

3.1.1 螺纹分类

一般情况下，螺纹按用途可分为连接螺纹（紧固螺纹）和传动螺纹、管螺纹和专用螺纹；按牙型可分为三角形螺纹、梯形螺纹、矩形螺纹、锯齿形螺纹和圆弧螺纹；按螺纹旋向可分为左旋螺纹和右旋螺纹；按螺旋线条数可分为单线螺纹和多线螺纹；按螺纹母体形状可分为圆柱螺纹和圆锥螺纹等。

3.1.2 公制螺纹的表示方法

ISO 标准公制螺纹牙型角为 60°，可分为粗牙和细牙两种，如图 3-2 所示，其各部分的尺寸表示及各参数的含义如下。

P：螺距。

β：牙型角。

d_1：外螺纹小径。

D_1：内螺纹小径。

d_2：外螺纹中径。

D_2：内螺纹中径。

d：外螺纹大径。

D：内螺纹大径。

φ：螺纹的螺旋升角。

螺纹的标注主要由螺纹代号、公称直径 $d(D)$、螺距 P、公差带代号和旋合长度组成。普通公制螺纹代号用字母 M 表示。常见的螺纹通常为右旋螺纹，无须特殊标注；如果是左旋螺纹，则需要标注 LH。如果没有标注螺距 P，则表示为粗牙螺纹，细牙螺纹需要标注螺距。

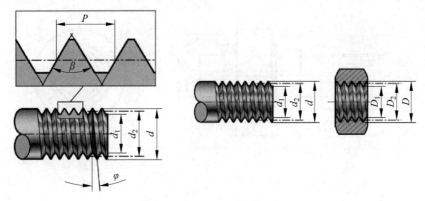

图 3-2 公制螺纹的尺寸表示法

例如,M24×1.5-6g 表示公称直径为 24mm,螺距为 1.5mm,公差带代号为 6g 的公制细牙普通螺纹。

3.1.3 螺纹车削相关尺寸计算

外螺纹大径 $d=$ 公称直径,在实际加工时,由于外螺纹的公差主要为下偏差,所以通常将外螺纹大径加工至 $d-0.1P$,外螺纹小径 $d_1=D-1.3P$。

3.2 螺纹检测量具

外螺纹主要用螺纹环规(见图 3-3)进行检测,内螺纹用螺纹塞规进行检测,两者都属于极限量规,在生产中广泛应用。采用该量具可以快速检测零件是否合格,但是不能测量出其具体尺寸数值。

注意:在检测之前,要看清螺纹环规标识,包括公称直径、螺距及公差带代号等。另外,还需清理干净被测

图 3-3 螺纹环规

螺纹油污及杂质,然后在螺纹环规与被测螺纹对正后,用大拇指与食指转动螺纹环规,使其在自由状态下旋合。若通过螺纹全部长度,则判定为合格;若未通过螺纹全部长度,则判定为不合格。

3.3 基本指令

螺纹加工循环指令 G92 主要用于等螺距的直螺纹和锥螺纹切削加工。G92 指令的指令格式为:

G92 X(U)__ Z(W)__ R__ F__;

其中,X(U)__ Z(W)__ 表示螺纹的终点坐标;R 表示锥螺纹起点与终点在 X 轴方向的坐标增量(半径值),在圆柱螺纹切削循环时,R 为"0",可省略;F 表示螺纹的导程 L,对于单线螺纹,导程=螺距 P。

G92 指令的指令动作如图 3-4 所示,刀具首先从循环起点沿着 X 轴快速进刀,按照编

制轨迹切削加工,退刀,快速返回。在执行指令过程中,需要注意以下两点。

(1) 在执行螺纹加工循环指令过程中,主轴和进给倍率无效。

(2) 如果在螺纹切削中(第二个动作)应用进给暂停,则刀具立即一边执行倒棱处理一边退刀,并按照平面第二轴(X 轴)、平面第一轴(Z 轴)的顺序返回到起点。

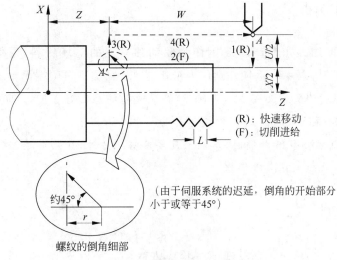

(a) 直螺纹加工

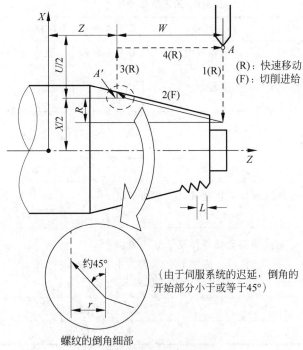

(b) 锥螺纹加工

图 3-4　螺纹加工循环指令

任务二 工艺准备

3.4 零件图分析

根据零件的使用要求,选择 45 钢作为驱动轴零件的毛坯材料,毛坯下料尺寸定为 $\phi50\times95$。在加工时,以 $\phi50$ 毛坯外圆作为粗基准,粗、精加工右侧端面、$\phi20$ 和 $\phi45$ 圆柱表面至要求尺寸,然后掉头装夹在 $\phi20$ 外圆处(在装夹时注意做好保护,以防表面夹伤),加工零件左端 $\phi25$、$\phi30$ 和 $M20\times1.5$ 外螺纹至要求尺寸。

注意:在车削右侧 $\phi25$ 外圆时,车削长度要足够。另外,在装夹毛坯时,应注意棒料伸出的长度,以免刀具与卡盘发生碰撞。

3.5 工艺设计

根据零件图分析,确定工艺过程,如表 3-1 所示。

表 3-1 工艺过程卡片

机械加工工艺过程卡片		产品型号	CLJG-01	零部件序号	QDZ-01	第 1 页	
		产品名称	槽轮机构	零部件名称	驱动轴	共 1 页	
材料牌号	C45	毛坯规格	$\phi50\times95$	毛坯质量	kg	数量	1
工序号	工序名	工序内容		工段	工艺装备	工时/min	
						准结	单件
5	备料	按 $\phi50\times95$ 尺寸备料		外购	锯床		
10	车加工	以 $\phi50$ 毛坯外圆作为粗基准,粗、精加工驱动轴零件		车	车床外径千分尺	45	30
15	车加工	以已加工完成的驱动轴零件及 $\phi20$ 外圆作为精基准(尽量同轴),粗、精加工 $\phi25$、$\phi30$ 外圆表面及 $M20\times1.5$ 外螺纹,并保证长度要求		车	车床外径千分尺	45	30
20	清理	清理工件,锐角倒钝		钳			5
25	检验	检验工件尺寸		检			5

本训练任务针对第 10、第 15 工序车削加工,进行工序设计,制订工序卡片,如表 3-2 所示。

表 3-2 车削加工工序卡片

机械加工工序卡片	产品型号	CLJG-01	产品型号	CLJG-01	第 1 页
	产品名称	槽轮机构	零部件名称	驱动轴	共 1 页

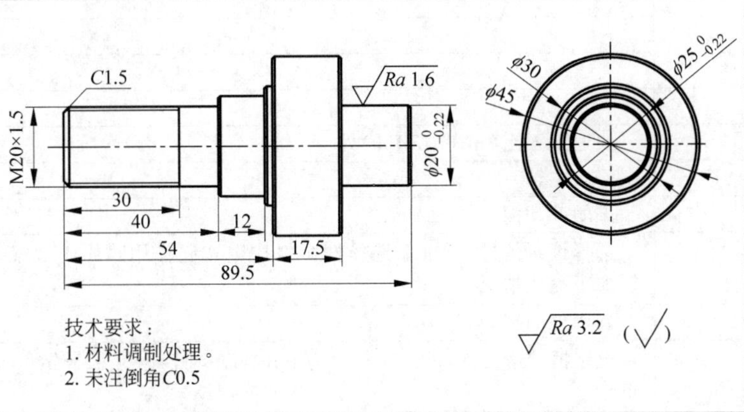

			工序号	10、15	
			工序名	车加工	
			材料	C45	
			设备	数控车床	
			设备型号	CK6150e	
			夹具	三爪自定心卡盘	
			量具	游标卡尺	
				千分尺	
			准结工时	90min	
			单件工时	60min	

技术要求：
1. 材料调制处理。
2. 未注倒角C0.5

工步	工步内容	刀具	S/(r/min)	F/(mm/r)	a_p/mm	工步工时/min 机动	工步工时/min 辅助
1	工件安装						5
2	粗加工 φ20 和 φ45 外圆表面及端面，精加工余量为 0.2mm	外圆粗车刀	1200	0.2	1.5	15	
3	精加工 φ20 和 φ45 外圆表面及端面	外圆精车刀	1500	0.1	0.2	10	
4	粗加工 φ20、φ25 和 φ30 外圆表面及端面，精加工余量为 0.2mm	切断刀（刀宽3mm）	700	0.07	0.1	15	
5	加工 M20×1.5 外螺纹	外螺纹车刀	500	1.5		10	5
6	拆卸、清理工件						5

3.6 数控加工程序编写

根据工序加工工艺，分别编写右端加工程序和左端加工程序，如表 3-3 和表 3-4 所示。

表 3-3 驱动轴右端数控加工程序

序号	程序语句	注 解
	O0001；	右侧数控加工程序
N1	T0101；	调用外圆车刀
	G97 G99 S800 M03；	设定恒转速控制，进给量单位为 mm/r，主轴转速为 800r/min，正转
	G0 X52 Z2 M8；	快速定位到循环起点（X52,Z2）处，打开冷却液
	G71 U1.5 R0.5；	调用纵向粗加工循环指令
	G71 P10 Q20 U0.4 W0 F0.2；	径向精加工余量为 0.4mm（单边0.2mm），轴向精加工余量为 0.05mm，进给量为 0.2mm/r
N10	G0 X0；	
	G1 G42 Z0；	

续表

序号	程序语句	注解
	X20 C0.5；	
	Z-18；	
	X45 C0.5；	
	Z-40；	
N20	G1 G40 X52；	
	G0 X100 Z150；	快速退刀至(X100,Z150)处
	M5；	停转主轴
	M9；	关闭冷却液
	M01；	选择性暂停(需要按下选择性暂停按键才起作用)，用于观察粗加工完成情况
N2	T00101；	调用外圆车刀
	G97 G99 S900 M3；	设定恒转速控制，进给量单位为mm/r，主轴转速为900r/min，正转
	G0 X52 Z2 M8；	快速定位到循环起点(X32,Z2)处，打开冷却液
	G70 P10 Q20 F0.1；	调用精加工循环指令，进给量为0.1mm/r
	G0 X100 Z150；	快速退刀至(X100,Z150)处
	M5；	停转主轴
	M9；	关闭冷却液
	M30；	程序结束

表3-4 驱动轴左端数控加工程序

序号	程序语句	注解
	O0002；	**左侧数控加工程序**
N1	T0101；	调用外圆车刀
	G97 G99 S900 M3；	设定恒转速控制，进给量单位为mm/r，主轴转速为900r/min，正转
	G0 X52 Z10；	快速定位到(X52,Z10)处
	G71 U1.5 R0.5；	调用纵向粗加工循环指令，背吃刀量为1.5mm，退刀量为0.5mm
	G71 P30 Q40 U0.4 W0.05 F0.18；	径向精加工余量为0.4mm(单边0.2mm)，轴向精加工余量为0.05mm，进给量为0.18mm/r
N30	G0 X0；	
	G1 G42 Z0；	
	X19.85 C1.5；	
	Z-40；	
	X30 C0.5；	
	Z-52；	
	X35 C0.5；	
	Z-54；	
	X44；	
	X46 Z-55；	
N40	G1 G40 X52；	

续表

序号	程序语句	注　　解
	G0 X100；	径向退刀
	Z150；	轴向退刀
	M5；	停转主轴
	M9；	关闭冷却液
	M01；	选择性暂停(需要按下选择性暂停按键才起作用)，用于观察粗加工完成情况
N2	T00101；	调用外圆车刀
	G97 G99 S900 M3；	设定恒转速控制，进给量单位为 mm/r，主轴转速为 900r/min，正转
	G0 X52 Z2 M8；	快速定位到循环起点(X52,Z2)处，打开冷却液
	G70 P30 Q40 F0.1；	调用精加工循环指令，进给量为 0.1mm/r
	G0 X100 Z150；	快速退刀至(X100,Z150)处
	M5；	停转主轴
	M9；	关闭冷却液
	M01；	选择性暂停(需要按下选择性暂停按键才起作用)，用于观察粗加工完成情况
N3	T0202；	调用外螺纹车刀
	G97 G99 S500 M3；	设定恒转速控制，进给量单位为 mm/r，主轴转速为 500r/min，正转
	G0 X22 Z5 M8；	快速定位到循环起点(X22,Z5)处，打开冷却液
	G92 X19 Z-30 F1.5；	调用螺纹加工循环指令 G92，第一次切削，进给量 F 为 1.5mm/min
	X18.4；	第二次切削
	X18.2；	第三次切削
	X18.05；	第四次切削
	X18.05；	光整牙型表面
	G0 X100 Z150；	快速退刀至(X100,Z150)处
	M5；	停转主轴
	M9；	关闭冷却液
	M30；	程序结束

任务三　上机训练

3.7　设备与用具

设备：CK6150e 数控车床。

刀具：外圆车刀、切断刀(刀宽 3mm)。

夹具：三爪自定心卡盘。

工具：卡盘扳手、刀架扳手等。

量具：0～150mm 游标卡尺、0～25mm 外径千分尺。

毛坯：$\phi 50 \times 95$。

辅助用品：垫刀片、毛刷等。

3.8 开机前检查

参考表 3-5 对机床状态进行点检。

表 3-5 机床开机准备卡片

检查项目		检查结果	异常描述
机械部分	主轴部分		
	进给部分		
	刀架		
	三爪自定心卡盘		
电器部分	主电源		
	冷却风扇		
数控系统	电气元件		
	控制部分		
	驱动部分		
辅助部分	冷却系统		
	压缩空气		
	润滑系统		

3.9 加工前准备

在加工前，应先将本任务所需刀具准备齐全，并安装正确。根据工艺要求设定工件原点，录入数控加工程序，并进行图形校验。

3.10 零件加工

在图形校验过程验证无问题后，即可进行零件加工。在零件加工前，应详细了解机床的安全操作要求，穿戴好劳动保护服装和用具。在进行零件加工时，应熟悉数控车床各操作按键的功能和位置，了解紧急状况的处置方法。在加工过程中，尤其是在即将切削之前，应对照显示屏"余移动量"栏显示的剩余移动量，观察刀具与工件之间的实际距离。若实际距离与剩余移动量相差过大，则应果断停机检查，以免发生撞机事故。若有异常，则应及时停止机床运动。

3.11 零件检测

在零件加工完成后，应当认真清理工件，并按照质量管理的相关要求，对加工完成的零件进行相关检验，保证生产质量。机械加工零件"三级"检验卡片如表 3-6 所示。

表 3-6 机械加工零件"三级"检验卡片

零部件图号		零部件名称		工 序 号	
材料		送检日期		工序名称	
检验项目	自检结果	互检结果	专业检验	备注	
检验结论	□合格　□不合格　□返修　□让步接收　检验签章：　　　　年　　月　　日				
不符合项描述					

项 目 总 结

驱动轴作为数控车床的典型加工零件，在生产和生活中应用广泛。根据设备情况和精度要求，其加工工艺也会存在一些差别。编程人员及操作人员需要结合加工条件，合理制定加工工艺，以提高零件的加工精度和生产效率。

课 后 习 题

1. 填空题

（1）螺纹按用途可分为_____和_____、_____和_____。

（2）在 FANUC 系统数控车床内、外圆切削单一固定循环采用_____指令来指定，而端面切削循环则采用_____指令来指定。

（3）如果没有标注螺距 P，则表示为_____，_____需要标注螺距。

（4）G92 指令主要用于等螺距的_____和_____切削加工。

（5）我们常常查表确定螺纹加工的进刀量，且进刀量需要遵循_____原则。

2. 判断题

（1）螺纹加工循环指令 G92 只能用于车削直螺纹，而不能车削锥螺纹。　　（　　）

（2）在 G92 指令执行过程中，机床面板上的进给速度倍率旋钮和主轴速度倍率旋钮均无效。　　（　　）

（3）G92 指令为模态代码。　　（　　）

(4) 在螺纹检测量具的选择上，外螺纹主要用螺纹环规检测，内螺纹用螺纹塞规检测。
（   ）

(5) G92 指令的指令格式中"F"表示的是进给速度。 （   ）

3. 选择题

(1) 加工表面多而复杂的零件，工序划分常采用（   ）。
    A. 按所用刀具划分　　　　　　B. 按安装次数划分
    C. 按加工部分划分　　　　　　D. 按粗精加工划分

(2) 在确定数控机床坐标轴时，一般应先确定（   ）。
    A. X 轴　　　B. Y 轴　　　C. Z 轴　　　D. A 轴

(3) 下列叙述中错误的是（   ）。
    A. 每个存储在系统中的数控程序必须制定一个程序号
    B. 程序段由一个或多个指令构成，表示数控机床的全部动作
    C. 在大部分系统中，程序段号仅作为"跳转"或"程序检索"的目标位置指示
    D. FANUC 系统的程序注释用"()"括起来

(4) 以下选项不是用于螺纹加工的指令是（   ）。
    A. G32　　　B. G90　　　C. G92　　　D. G71

(5) 用 FANUC 系统指令"G92 X(U)__ Z(W)__ F __;"加工双线螺纹，则该指令中的 "F"是指（   ）。
    A. 螺纹导程　　B. 螺纹螺距　　C. 每分钟进给量　D. 螺纹起始角

4. 简答题

(1) 简述螺纹的分类。

(2) 简述螺纹的测量方法。

(3) 简述螺纹加工循环指令 G92 的主要应用场合。

5. 综合编程题

根据图 3-5 所示的驱动轴图纸，按照要求编写驱动轴数控加工程序，并完成零件加工。自我学习检测评分表如表 3-7 所示。

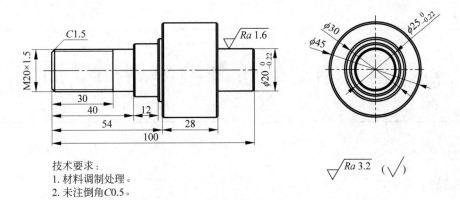

技术要求:
1. 材料调制处理。
2. 未注倒角C0.5。

图 3-5　题 5 图

表 3-7　自我学习检测评分表

项　目	目标要求	分值	评分细则	得分	备注
学习关键知识点	(1) 了解螺纹轴的分类及其加工特点 (2) 能正确使用螺纹检测量具 (3) 掌握螺纹加工循环指令的使用,理解各参数的含义	20	理解与掌握		
工艺准备	(1) 能够正确识读零件图 (2) 能够独立确定加工工艺路线,并正确填写工艺文件 (3) 能根据工序加工工艺,编写正确的加工程序	30	理解与掌握		
上机训练	(1) 能够根据零件结构特点和精度合理选用量具,并正确、规范地测量出相关尺寸 (2) 掌握驱动轴车削加工的操作流程 (3) 能够正确操作数控车床,并根据加工情况调整加工参数	50	(1) 理解与掌握 (2) 操作流程		

思政小课堂

项目四　驱动轮铣削编程与加工训练

> **思维导图**

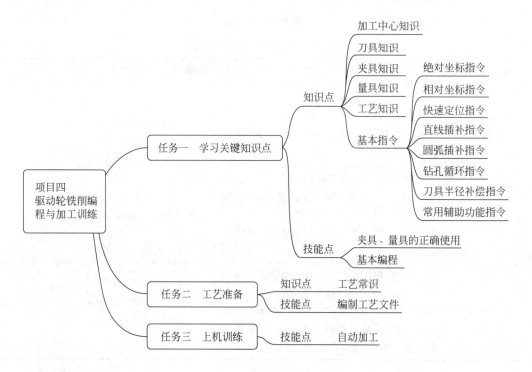

> **学习目标**

知识目标

（1）具备基本平面类零件的识图能力。
（2）了解驱动轮在本机构中的用途和关键要求。

能力目标

（1）掌握基本准备指令和辅助指令的使用方法。
（2）掌握铣削刀具的选择与安装方法。
（3）能够独立确定加工工艺路线，并正确填写工艺文件。
（4）能够正确操作数控加工中心，并根据加工情况调整加工参数。
（5）能够根据零件结构特点和精度合理选用量具，并正确、规范地测量相关尺寸。

素养目标

（1）培养学生的科学探究精神和态度。
（2）培养学生的工程意识。
（3）培养学生的团队合作能力。

任务一　学习关键知识点

4.1　初步认识数控加工中心

数控铣床是用电子数字化信号控制的铣床。数控铣床是在一般铣床的基础上发展起来的一种自动加工设备，两者的加工工艺基本相同、结构类似。数控铣床分为不带刀库和带刀库两大类。其中，带刀库的数控铣床又称加工中心。

工件在加工中心上经一次装夹后，数控系统能控制机床按不同工序，自动选择和更换刀具，自动改变机床主轴转速、进给量和刀具相对工件的运动轨迹及其他辅助机能，依次完成工件几个面上多工序的加工。并且，加工中心有多种换刀及选刀功能，从而使生产效率大幅度提高。

立式加工中心是指主轴轴线与工作台垂直设置的加工中心，主要适用于加工板类、盘类、模具及小型壳体类复杂零件。立式加工中心能完成铣削、镗削、钻削、攻螺纹和切削螺纹等工序。立式加工中心最少是三轴二连动，一般可实现三轴三连动。有的可进行五轴、六轴控制。立式加工中心立柱高度有限，相应地，对箱体类工件的加工范围比较小，这是立式加工中心的缺点。但立式加工中心工件装夹、定位方便；刃具运动轨迹易观察，调试程序检查测量方便，可及时发现问题，进行停机处理或修改；冷却条件易建立，冷却液能直接到达刀具和加工表面；三个坐标轴与笛卡儿坐标系吻合，直观感觉与图样视角一致；切屑易排除、易掉落，避免划伤加工过的表面。与相对应的卧式加工中心相比，其结构简单、占地面积较小、价格更低。

AVL650e立式加工中心（见图4-1）立柱底部为A字形高刚性结构设计，床身采用高强度高级铸铁结构，各轴滑轨全支撑；三轴滑轨采用线性滚动导轨，采用C3级精密滚珠丝杆，定位精度佳；高刚性高速主轴，转速可达10000r/min，主电机以H.T.D齿形皮带直接带动主轴运动；刀臂式自动换刀装置，刀库容量为24把刀，换刀时间仅为2.5s（刀对刀）；标配螺旋式排屑机及集屑车，排屑易，减少铁屑清除辅助时间；电气箱配备热交换器，有效降低电气元件故障率，提高机床使用寿命；为保证加工精度，机床标配主轴油温冷却装置。

图4-1　AVL650e立式加工中心

AVL650e立式加工中心主要性能参数如表4-1所示。

表 4-1 AVL650e 立式加工中心主要性能参数

项目内容	技术规格
X/Y/Z 行程/mm	620/520/520
主轴鼻端至工作台面距离/mm	100～620
主轴中心至立柱滑轨面距离/mm	540
工作台尺寸/(mm×mm)	800×500
工作台最大载重/kg	500
T 形槽尺寸(槽宽/槽距/槽数)/(mm×mm×mm)	18×130×3
主轴转速/(r/min)	100～10000
主轴锥度	BT40
快速进度速度(X/Y/Z)/(m/min)	48/48/48
切削进给速度(X/Y/Z)/(m/min)	1～20000
刀库容量/pcs	24
换刀方式	刀臂式
换刀时间(刀对刀)/s	2.5
电源容量/kV·A	20
机床尺寸/(mm×mm×mm)	2400×2300×2700
机床质量/kg	4200
定位精度/mm	0.01/全长
重复定位精度/mm	0.005
气压要求/bar①	6
数控系统	FANUC 0i-MF

4.2 铣削加工特点

铣削是以铣刀的旋转运动为主运动,以铣刀或工件进给运动的一种切削加工方法。其特点如下。

(1) 采用多刃刀具加工,刀刃轮替切削,刀具冷却效果好、耐用度高。但与车削相比,铣削的切削过程不连续,切削层参数及切削力是变化的,容易引起冲击和振动,从而影响加工质量的提高。

(2) 铣削加工生产效率高、加工范围广,在普通铣床上使用各种不同的铣刀可以完成加工平面(平行面、垂直面、斜面)、台阶、沟槽(直角沟槽、V 形槽、T 形槽、燕尾槽等特形槽)、特形面等的加工任务。加上分度头等铣床附件的配合运用,还可以完成花键轴、螺旋轴、齿式离合器等工件的铣削。

(3) 铣削加工具有较高的加工精度,其经济加工精度一般为 IT9～IT7,表面粗糙度 Ra 一般为 12.5～1.6μm。精细铣削精度可达 IT5,表面粗糙度 Ra 可达到 0.20μm。

与车削加工不同,铣削加工参数的定义略有不同。

如图 4-2 所示,在铣削加工中,切削速度通常用主轴转速来表示,主轴转速 n 是铣刀在主轴上每分钟的转数,单位为 r/min;进给速度 F 通常用工件与刀具中心每分钟的相对移

---

① 1bar=100kPa。

动的路程来表示,单位为 mm/min;切削深度可分为轴向切深 a_p 和径向切深 a_e,分别表示刀具在工件表面的金属去除量和铣刀直径沿径向参与零件切削的宽度,单位为 mm。

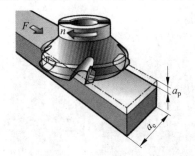

图 4-2　铣削加工模型

4.3　刀具知识

铣削加工刀具种类很多,一般按加工特点可分为:面铣刀(平铣刀)、肩铣刀(侧铣刀)、仿形铣刀、槽铣刀、螺纹铣刀、倒角铣刀和齿轮铣刀等,如图 4-3 所示。

(a) 面铣刀　　(b) 肩铣刀　　(c) 仿形铣刀　　(d) 槽铣刀

(e) 螺纹铣刀　　(f) 倒角铣刀　　(g) 齿轮铣刀

图 4-3　常用用途的铣刀

常用铣刀的结构有整体铣刀、焊片铣刀和嵌片铣刀,如图 4-4 所示。整体铣刀从本体至刀刃末端由同一种材质制成,大多为高速钢、硬质合金、金属陶瓷材质,多数都是直径较小或特殊形状的铣刀;焊片铣刀刀柄部分及本体采用价格较低的碳素工具钢材料,刃部钎焊高速钢、碳化物、立方氮化硼(CBN)、金刚石等;嵌片铣刀采用螺钉或夹片将刀片固定在刀体上。

(a) 整体铣刀　　　　(b) 焊片铣刀　　　　(c) 嵌片铣刀

图 4-4　常用铣刀结构

在数控铣削刀具的作用过程中,应注意刀柄在机床主轴上的安装方式,即刀柄的类型。刀柄的不当使用可能会损坏刀柄,甚至机床主轴。在日常工作中,加工中心常用的刀柄类型如图4-5所示。

图 4-5　加工中心常用的刀柄类型

刀柄是一种工具,是机械主轴与刀具和其他附件工具的连接件,主要标准有BT、JT(SK)、CAPTO、BBT、HSK等几种规格。

BT刀柄是加工中心主轴采用7∶24锥度实现刀具与机床主轴的连接。标准的7∶24锥面连接有许多优点,例如,可实现快速装卸刀具;刀柄的锥体在拉杆轴向拉力的作用下,紧紧地与主轴的内锥面接触,实心的锥体直接在主轴的锥孔内支撑刀具,可减小刀具的悬伸量;只有一个尺寸且加工精度高,所以成本较低且可靠。

JT刀柄相当于BT刀柄的进化版,在德国也称SK刀柄。JT刀柄比BT刀柄多了一个和主轴的结合面,但定位的缺口对于在高转速情况下的动平衡性能不利,需要通过去重等方法提高动平衡性能。JT刀柄与BT刀柄的不同之处:刀柄与主轴端面键配合的键槽不同,BT刀柄的键槽是半截的,而JT刀柄是通槽;JT刀柄的抓刀环上有一个豁口,BT刀柄的抓刀环上则没有;两者抓刀环部分的公称尺寸不同,换刀机械手爪的薄厚不一样。所以,在使用过程中,两者不可互换。

HSK刀柄是高速切削应用刀具的刀柄,高速切削加工已成为机械加工制造技术的重要环节。HSK刀柄是一种新型的高速短锥型刀柄,其接口采用锥面和端面同时定位的方式,刀柄为中空,锥体长度较短,锥度为1∶10,有利于实现换刀轻型化和高速化。由于采用空心锥体和端面定位,其补偿了高速加工时主轴孔与刀柄的径向形变差异,并完全消除了轴向定位误差,使高速、高精度加工成为可能。这种刀柄在高速加工中心上应用越来越广泛。

4.4　夹具知识

夹具是指在机械制造过程中用来固定加工对象,使其占有正确的位置,以接受施工或检测的装置,又称卡具。从广义上说,在工艺过程中的任何工序中,用以迅速、方便、安全地安装工件的装置,都可称为夹具。夹具通常由定位元件(确定工件在夹具中的正确位置)、夹紧装置、对刀引导元件(确定刀具与工件的相对位置或导引刀具方向)、分度装置(使工件在一次安装中能够完成数个工位的加工,分为回转分度装置和直线移动分度装置两类)、连接元件及夹具体(夹具底座)等组成。

铣床夹具主要用于加工零件上的平面、凹槽、花键及各种成型面,是最常用的夹具之一。主要由定位元件、夹紧装置、夹具体、连接元件、对刀引导元件组成。在铣削加工时,切削力度较大,又是断续切削,振动较大,因此,铣床夹具的夹紧力要求较大,夹具刚度、强度的要求都比较高。

与其他机床夹具系统相同,铣床夹具分为通用夹具、专用夹具、可调夹具、组合夹具等类型,按夹紧方式又分为手动夹具、气动夹具、液压夹具、电动夹具、磁力夹具和真空夹具等。图 4-6 为加工中心常用的手动通用夹具。

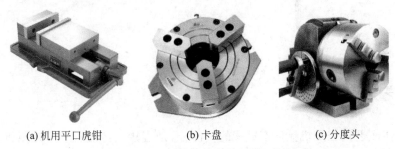

(a) 机用平口虎钳　　(b) 卡盘　　(c) 分度头

图 4-6　加工中心常用的手动通用夹具

机用平口虎钳是一种机床通用附件,配合工作台使用,对加工过程中的工件起固定、夹紧、定位作用。由躯座、活动钳口、螺母、螺杆等构件组成,按其结构和使用可分为通用平口虎钳、角度压紧机用平口虎钳、可倾机用平口虎钳、高精度机用平口虎钳、增力机用平口虎钳等。

卡盘是机床上用来夹紧工件的机械装置,其利用均布在卡盘体上的活动卡爪的径向移动,将工件夹紧并定位。卡盘一般由卡盘体、活动卡爪和卡爪驱动机构 3 部分组成。卡盘体直径最小为 65mm,最大可达 1500mm,中央有通孔,以便通过工件或棒料;其背部有圆柱形或短锥形结构,可直接通过法兰盘与机床相连接,也可通过压板压紧在机床工作台上。

分度头是安装在铣床上的用于将工件分成任意等份的机床附件。其利用分度刻度环和游标、定位销、分度盘及交换齿轮,将装卡在顶尖或卡盘上的工件分成任意角度,且可将圆周分成任意等份,辅助机床利用各种不同形状的刀具进行各种沟槽、正齿轮、螺旋正齿轮、阿基米德螺线凸轮等的加工工作。在数控机床上,数控分度头采用 AC 或 DC 伺服电机进行驱动,利用复节距蜗杆蜗轮组机构进行传动,使用油压环抱式锁紧装置,且配备扎实的刚性密封结构。数控分度头广泛适用于铣床、钻床及加工中心。配合工作母机四轴操作界面,可进行同动四轴加工。

4.5　量具知识

在铣床安装夹具或工件时,还经常需要保证夹具或工件的某一基准与机床的坐标轴平行,即拉直,此时需要用到百分表,如图 4-7 所示。在一些模具零件加工前,往往也需要使用百分表找正工件坐标系。

杠杆百分表又称杠杆表或靠表,是利用杠杆—齿轮传动机构或杠杆—螺旋传动机构,将尺寸变化为指针角位移,并指示出长度尺寸数值的计量器具。其用于测量工件几何形状误差和相互位置的正确性,并可用比较法测量长度。

图 4-7 百分表

4.6 工艺知识

机械加工工艺过程是机械生产过程的一部分,是直接生产过程。它是用机械加工的方法,直接改变毛坯的形状、尺寸和表面质量,使其成为合格产品零件的过程。

4.6.1 定位原理

在机械加工中,工件的尺寸、形状和表面之间的位置精度是由刀具和工件的相对位置来保证的。在加工前,将工件装在相对刀具的一定位置上,称为定位。

6点定位法则。任何一个未被约束的物体,在空间中有 6 个自由度。要使物体在空间中有确定的位置,必须约束这 6 个自由度。工件定位的实质是使工件在夹具中占据确定的位置,因此,工件的定位问题可转化为在空间直角坐标系中决定刚体坐标位置的问题来讨论。如图 4-8 所示,在空间直角坐标系中,刚体具有 6 个自由度,即沿 X 轴、Y 轴、Z 轴移动的 3 个自由度和绕此三轴旋转的 3 个自由度。用 6 个合理分布的支承点限制工件的 6 个自由度,使工件在夹具中占据正确的位置,称为 6 点定位法则。人们在阐述 6 点定位法则时,常以图 4-8 所示的铣不通槽的示例来加以说明。a_1、a_2、a_3 这 3 个点体现主定位面 A,限制 X、Y 方向的旋转自由度和 Z 方向的移动自由度;a_4、a_5 两个点体现侧面 B,限制 X 方向的移动自由度和 Z 方向的旋转自由度;a_6 点体现止推面 C,限制 Y 方向的移动自由度。这样,工件的 6 个自由度全部被限制,称为完全定位。当然,定位只是保证工件在夹具中的位置确定,并不能保证在加工中工件不移动,故还需夹紧。定位和夹紧是两个不同的概念。

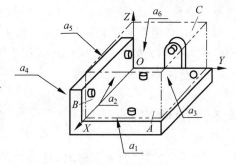

图 4-8 6 点定位原理

在生产应用中,有如下 4 种定位形式。

(1) 完全定位。工件的定位采用了 6 个支承点,限制了工件全部的 6 个自由度,使工件在夹具中有唯一确定的位置。

(2) 不完全定位。根据加工需要,只限制了工件的部分自由度。

(3) 欠定位。根据加工技术要求,没有限制应该限制的自由度的定位。

(4) 过定位。工件的同一自由度被两个及以上的不同定位元件重复限制的定位。

4.6.2 基准

在零件的设计和制造过程中,要确定一些点、线或面的位置,必须以一些指定的点、线或面作为依据,这些作为依据的点、线或面,称为基准。按照作用的不同,常把基准分为设计基准和工艺基准两类。

设计基准是指设计零件的基准。

工艺基准是指在制造零件时所使用的基准,它又分为工序基准、定位基准、测量基准、装配基准。工序基准是指在工艺文件上用以标定加工表面位置的基准;定位基准是指在机械加工中,用来使工件在机床或夹具中占有正确位置的点、线或面,它是工艺中最主要的基准,其选择是否合理,对保证工件加工后的尺寸精度和形位精度、安排加工顺序、提高生产率,以及降低生产成本起着决定性的作用,它是制定工艺过程的主要任务之一,可分为粗基准和精基准两种;测量基准是指用以测量已加工表面尺寸及位置的基准;装配基准是指用来确定零件或部件在机器中的位置的基准。

选择定位基准是为了保证工件的位置精度,因此,定位基准总是从有位置精度要求的表面开始进行选择的。粗基准是毛坯表面的定位基准,在选取时一般遵循以下原则。

(1) 选取不加工的表面作为粗基准,这样可使加工表面具有较正确的相对位置,并有可能在一次安装中就可加工大部分加工表面。

(2) 选取要求加工余量均匀的表面作为粗基准,这样可以保证作为粗基准的表面在加工时余量均匀。

(3) 对于所有表面都要加工的零件,选取余量和公差最小的表面作为粗基准,以避免余量不足而造成废品。

(4) 选取光洁、平整、面积大的表面作为粗基准。

(5) 粗基准不应重复使用。一般情况下,粗基准只允许使用一次。

对于形位公差精度要求较高的零件,应采用已经加工过的表面作为定位基准。这种定位基准称为精基准。精基准的选择一般遵循以下原则。

(1) 基准重合原则:选用定位基准与设计基准重合的原则。如图 4-9(a)所示,在加工孔 D 时,为保证尺寸 A,选择 K 为精基准。

(2) 基准统一原则:位置精度要求较高的各加工表面,应尽可能在多数工序中统一用同一基准。如图 4-9(b)所示,在加工齿轮时,为保证精度,各精加工工序统一使用内孔 $\phi35$ 和端面 B 作为精基准。

(3) 互为基准原则:在需要加工的各表面中,加工时互相以对方为定位基准。

(4) 自为基准原则:以加工表面自身作为定位基准。

总之,无论是选择粗基准还是精基准,都必须首先使工件定位稳定、安全可靠,然后再考虑夹具设计容易、结构简单、成本低廉等技术、经济原则。

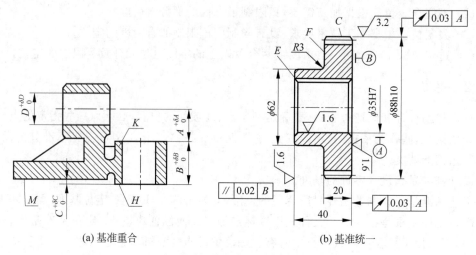

(a) 基准重合　　　　　(b) 基准统一

图 4-9　精基准选择原则

4.6.3　机械加工工艺规程制定的原则

在制定机械加工工艺规程时,应遵循以下原则。

(1) 基面先行原则。在零件加工时,必须选择合适的表面作为定位基准面,以便正确安装工件。在第一道工序中,只能用毛坯面(未加工面)作为定位基准面;在后续工序中,为了提高加工质量,应尽量采用加工过的表面为定位基准面。显然,在安排加工工序时,精基准面应先加工。

(2) 粗精分开,先粗后精原则。在对零件加工质量要求高时,对精度要求高的表面,应划分加工阶段。一般可分为粗加工、半精加工和精加工三个阶段,精加工应放在最后进行。这样,有利于保证加工质量,同时,也有利于某些热处理工序的安排。

(3) 先面后孔原则。对于箱体、支架类等零件,应先加工平面后加工孔。这是因为平面的轮廓平整,安放和定位稳定可靠。先加工好平面,就能以平面定位加工孔,保证平面和孔的位置。

综上所述,一般机械加工的顺序是:先加工精基准面→粗加工主要面(精度要求高的表面)→精加工主要面。为保证制造系统的能效,次要表面的加工可适当穿插在各阶段之间进行。

4.7　基本指令

4.7.1　数控铣削程序的基本格式

一个数控铣削程序的基本格式如表 4-2 所示。

程序头一般包括程序起始符和程序名或程序号,程序起始符根据不同的数控系统,可以为"％"":""MP"等不同的标识,有的系统也可以省略,在书写程序时需参照机床编程说明书进行指定。在 FANUC 0i 系统中,程序头可以直接以英文大写字母"O"后接数字的程序号来表示,程序号为正整数,取值范围为 0001～9999。在输入程序号时,数字前的 0 可以省略。

表 4-2 数控铣削程序基本格式

程序内容	程序步骤	程序语句	注 解
程序头	①	O0001;	程序名
程序体	②	T01 M6;	调用刀具
	③	S2000 M03;	主轴正转启动
	④	G0 G90 G54 X__ Y__ G43 Z__ H1;	在 G54 坐标系下移动到点（X__,Y__） Z 向移动到安全位置，启动刀具长度补偿
	⑤	G0 X__ Y__ Z__;	移动至下刀起点位置
	⑥	下刀、(进刀、切削、退刀)、抬刀	切削工件
	⑦	G0 Z__;	退回 Z 向安全位置
	⑧	M5;	主轴停转
	⑨		如需其他刀具加工，则重复步骤②~⑧
程序尾	⑩	M30;	程序结束

程序体为程序的主体部分，为刀具切削工件的过程。加工过程若只有一把刀，则程序体为步骤②~⑧；若有多把刀具，则重复步骤②~⑧过程。其中，步骤②~④为切削过程的初始化，在这些步骤中调用刀具、设置切削速度、选择工件坐标系和刀具补偿号、运动到安全位置，为正式切削加工作准备。步骤⑤~⑥为正式的切削过程，表示工件上一个区域的加工过程。若一个区域需要多次走刀切削，则只需重复"进刀、切削、退刀"的过程即可；若切削多个区域，则只需重复步骤⑤~⑥即可。在立式加工中心加工过程中，每把刀具在所有区域切削完成后，至少应将零件退回 Z 向安全位置，并用 M5 指令让主轴停转。

程序尾包括程序的结束指令和结束标识符。FANUC 系统常用的程序结束指令为 M02、M30。与程序头相似，结束标识符应根据机床编程说明书中的要求进行指定。

常用准备功能指令及辅助功能指令如表 4-3 所示。

表 4-3 常用准备功能指令、辅助功能指令表

指令	用 途	指令	用 途
G00	快速定位	G68	坐标系旋转（用 G69 指令取消）
G01	直线插补	G80	钻孔循环取消
G02	顺时针圆弧插补	G8*	G81~G87 孔加工循环
G03	逆时针圆弧插补	M00	程序暂停
G15	极坐标方式取消	M03	主轴正转
G16	极坐标方式切换	M04	主轴反转
G28	返回参考点	M05	主轴停转
G40	取消刀具半径补偿	M08	冷却液打开
G41	刀具半径左补偿	M09	冷却液关闭
G42	刀具半径右补偿	M30	程序结束，并返回程序头
G51	比例缩放（用 G50 指令取消）	M98	子程序调用
G52	局部坐标系	M99	子程序返回

4.7.2 常用准备功能指令

1. 绝对坐标指令与相对坐标指令

指定刀具的移动有两种方法，即绝对坐标指令和相对坐标指令。绝对坐标指令是对刀具的移动位置以工件坐标系实际坐标值的方法指定终点坐标；相对坐标指令也称增量坐标指令，是以刀具相对当前点以增量的形式指定终点坐标的方法。

如图 4-10 所示，在使刀具从起点移动到终点时，可用以下方法编写程序。

应用绝对坐标指令编写的程序为：

G90 X40.0 Y70.0;

应用相对坐标指令编写的程序为：

G91 X-60.0 Y40.0;

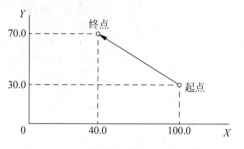

图 4-10　绝对坐标指令与相对坐标指令

2. 工件坐标系

工件坐标系是在编程时使用的坐标系，又称编程坐标系，该坐标系是为编程方便人为设定的。FANUC 系统可在 G54~G59 中指定 6 个工件坐标系，开机后默认的为 G54 坐标系，机床在操作时可通过 MDI 单元 Offset Setting 设置工件原点偏置值，还可通过程序方法或外部数据输入方法设置其偏置值。如图 4-11 所示，在应用中要注意机床的外部工件原点偏置值会影响 G54~G59 工件坐标系原点的位置。在加工中心的实际操作中，可通过机内工件找正的方法设置工件坐标系。

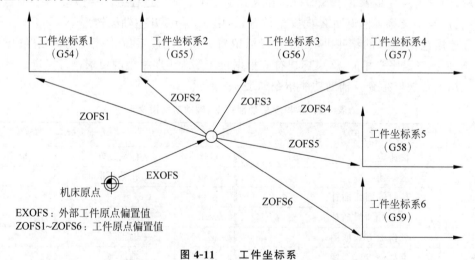

图 4-11　工件坐标系

3. 刀具长度补偿

如图 4-12 所示，通过在偏置存储器中设定编程时假想的刀具长度值与实际进行加工时使用的刀具长度值的偏移，不用更改程序就可以补偿刀具长度值的偏移。FANUC 系统通过 G43 指令指定偏置方向，再利用刀具长度补偿值指定地址之后的编号（H 代码），用以指定偏置存储器中设定的刀具长度补偿值。

在操作过程中,可以参考工件坐标系或机床坐标系建立刀具长度补偿,这个过程为机内对刀;也可以在工件找正时,确定机床主轴参考点与工件坐标系的偏置,而刀具通过机外对刀仪对刀,将刀具的实际长度导入机床。

4. 快速定位指令 G00

G00 指令是指将刀具从当前的定位点,以数控系统预设的最大进给速度,快速移动到程序段所指令的下一个定位点。如图 4-13 所示,快速定位过程可能存在多种运动轨迹,常见的轨迹为直线插补型定位和非直线插补型定位。由于机床在出厂时参数设定不同,轨迹也不同,在操作机床时应注意其运动轨迹,以免发生碰撞事故。快速定位指令格式为指令代码后接定位终点位置坐标:

G00 X__ Y__ Z__ ;

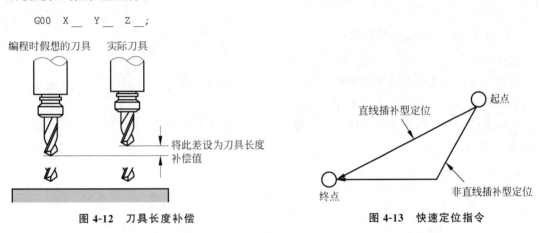

图 4-12 刀具长度补偿　　　图 4-13 快速定位指令

5. 直线插补指令 G01

G01 指令是指将刀具以直线形式按 F 代码指定的速率,从其当前位置移动到命令要求的位置。其特点是各坐标以联动的方式,按进给速度 F 作任意斜率的直线运动。其指令格式为:

G01 X__ Y__ Z__ F__ ;

如图 4-14 所示,刀具以 200mm/min 的速度从点 P_1 切削到点 P_2 用绝对坐标指令和相对坐标指令编程如下。

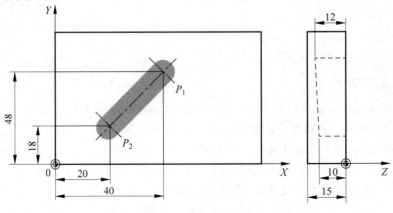

图 4-14 直线插补指令

(1) 绝对坐标编程：

G01 G90 X20 Y18 Z-10 F200;

(2) 相对坐标编程：

G01 G91 X-20 Y-20 Z2 F200;

6. 圆弧插补指令

G02 为顺时针圆弧插补指令，G03 为逆时针圆弧插补指令。在 XY 平面内常用的格式有两种。

(1) 指定半径法指令格式为：

G02/G03 X__ Y__ R__ F__;

(2) 指定圆心法指令格式为：

G02/G03 X__ Y__ I__ J__ F__;

指定半径法无法直接加工整圆。当所加工圆弧的圆心角不大于 180°时，R 为正值；当所加工圆弧的圆心大于 180°且小于 360°时，R 为负值。

指定圆心法可以用于加工整圆和其他任意圆弧，I 和 J 的值为起点指向圆心的矢量，如图 4-15 所示。

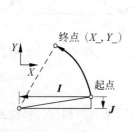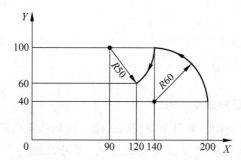

图 4-15 圆弧插补指令

(1) 指定半径法编程：

G03 X140 Y100 R60;
G02 X120 Y60 R50;

(2) 指定圆心法编程：

G03 X140 Y100 I-60 J0;
G02 X120 Y60 I-50 J0;

7. 钻孔循环指令

常用的钻孔循环指令有两种，一种为一般定位孔和浅孔，使用 G81 指令；另一种为加工深孔，使用 G83 指令。其指令格式分别如图 4-16 和图 4-17 所示。

其中，G98 指令表示在加工完一个孔后，抬刀到初始平面后再向下一个孔移动；G99 表示在加工完一个孔后，抬刀到点 R 坐标指定的平面后，再向下一个孔移动。在加工多个孔的实际操作中，要注意表面有无干涉，以免发生碰撞。

G81 X_Y_Z_R_F_K_;

X_Y_ ：孔位置数据
Z_ ：从R点到孔底的距离
R_ ：从基准平面到R点的距离
F_ ：切削进给速度
K_ ：重复次数（仅限需要重复时）

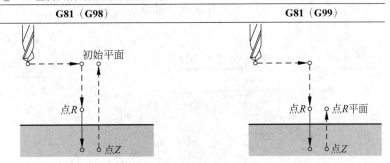

图 4-16　G81 指令钻孔循环

G83 X_Y_Z_R_Q_,D_F_K_;

X_Y_ ：孔位置数据
Z_ ：从R点到孔底的距离
R_ ：从基准平面到R点的距离
Q_ ：每次的进刀量
,D_ ：退刀量
F_ ：切削进给速度
K_ ：重复次数（仅限需要重复时）

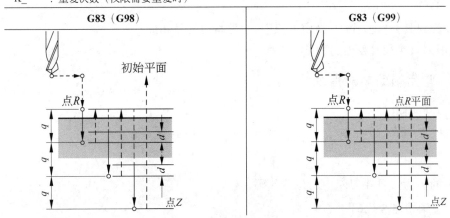

图 4-17　G83 指令钻孔循环

8．刀具半径补偿指令

　　数控机床在加工过程中，它所控制的是刀具中心的轨迹，而用户为了方便起见，总是按零件轮廓编制加工程序。因此，为了加工所需的零件轮廓，在进行内轮廓加工时，刀具中心必须向零件的内侧偏移一个刀具半径值；在进行外轮廓加工时，刀具中心必须向零件的外侧偏移一个刀具半径值。这种根据零件轮廓编制的程序和预先设定的偏置参数，数控装置实时自动生成刀具中心轨迹的功能称为刀具半径补偿功能。如图4-18(a)所示，实线为所需加工的零件轮廓，虚线为刀具中心轨迹。根据 ISO 标准，当刀具中心轨迹在编程轨迹（零件轮廓）前进方向的右边时，称为右刀补，用 G42 指令实现；反之称为左刀补，用 G41 指令实现。

如图 4-18(b)所示,为使偏置与刀具的半径值一样大,数控机床首先创建一个偏置矢量(起刀),其长度等于刀具半径。偏置矢量与刀具的前进方向垂直,即从工件朝向刀具中心的方向。如果在起刀后指定直线插补或圆弧插补,则可使刀具在仅偏置某一偏置矢量后进行加工。最后,为使刀具返回到起点,取消刀具半径补偿。

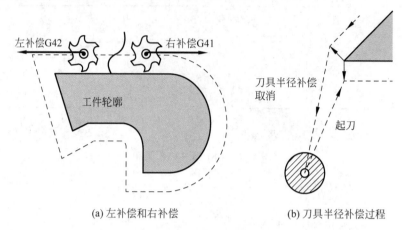

(a) 左补偿和右补偿　　(b) 刀具半径补偿过程

图 4-18　刀具半径补偿

在 XY 平面内,刀具半径补偿的指令格式为:

　　G1 G41/G42 X__ Y__ D__;

其中,D 为刀具补偿值代码,为便于识别,通常与刀具号相同。

取消刀具半径补偿时,使用 G40 指令。

4.7.3　常用辅助功能指令

1. 刀具调用

立式加工中心应用 M06 辅助指令加 T 代码调用刀具。指令格式为:

　　T__ M06;

其中,T 的值为刀具在刀库中的位置代码。

2. 主轴功能 S 代码

加工中心使用主轴功能 S 代码指定机床主轴的回转速度,单位为 r/min,当主轴功能 S 代码指定的数值为正整数,且数值范围大于机床的最大额定转速时,机床主轴以最大额定转速旋转。辅助功能 M 代码控制主轴的启停,M03 为主轴正转,M04 为主轴反转,M05 为主轴停转,通常将从 Z 轴负方向观察,主轴顺时针旋转定义为正转。其指令格式如下。

(1) 主轴正转指令格式为:

　　S__ M03;

(2) 主轴反转指令格式为:

　　S__ M04;

(3) 主轴停转指令格式为：

M05;

3. 程序结束指令

可以使用 M02 指令、M30 指令结束数控程序。其中，M02 指令表示结束主程序；M30 指令表示结束主程序，并复位。其具体意义与机床的出厂设置有关，可通过修改参数 No3404 ♯5 或 No3404 ♯4 重新定义。

任务二　工　艺　准　备

4.8　零件图分析

根据零件的使用要求，可以选择 45 钢作为驱动轮零件的毛坯材料，毛坯下料尺寸定为 $\phi 255 \times 25$。

如图 4-19 所示，$\phi 47$ 为主轴的安装尺寸，可通过车床精镗获得其精度，在加工时，以 $\phi 255$ 毛坯外圆和一个端面为粗基准，精加工内孔部分，因 $\phi 21$ 孔后续加工用于工件找正，故在加工时将其精度提高到 IT8(0.033mm)，然后以 $\phi 47$ 及精加工后的一端面为基准，加工 $\phi 250$、$\phi 160$，保证两处厚度为 10mm。

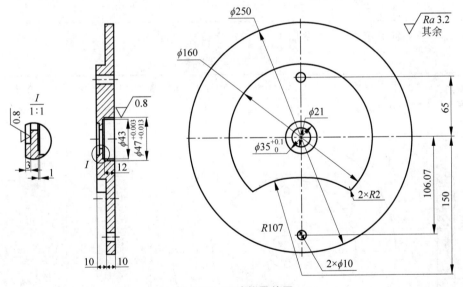

图 4-19　驱动轮零件图

在铣削加工时，直接以 $\phi 250$ 为基准，为保证精度，找正后可校准 $\phi 21$ 内孔，铣削 $R107$、$\phi 35$ 孔。因 $\phi 35$ 孔底有精度要求，故应采用粗加工→半精加工→精加工的顺序进行。$2 \times \phi 10$ 为拨销和手柄的安装孔，设计图采用螺纹连接，故精度要求不高，可直接通过钻孔获得。

注意： 此零件为手动操作件，所以在加工完成后，应去除加工毛刺，保证锐角充分倒钝，以确保在使用过程中的人身安全。

4.9 工艺设计

根据零件图分析,确定工艺过程,如表 4-4 所示。

表 4-4 工艺过程卡片

机械加工 工艺过程卡片	产品型号	CLJG-01	零部件序号	QDL-01	第 1 页	
	产品名称	槽轮机构	零部件名称	驱动轮	共 1 页	
材料 牌号	C45	毛坯规格	$\phi255\times25$	毛坯质量	10kg	数量	1

工序号	工序名	工序内容	工段	工艺装备	工时/min	
					准结	单件
5	备料	按 $\phi255\times25$ 尺寸备料	外购	锯床		
10	车加工	以 $\phi255$ 毛坯外圆和一个端面为粗基准,精加工 $\phi47$ 内孔部分,在加工时将 $\phi21$ 孔精度提高到 IT8(0.033mm)	车	车床内径百分表	60	45
15	车加工	以 $\phi47$ 内孔定位,加工外圆 $\phi250$;加工台阶 $\phi160$ 保证厚度为 10mm	车	车床游标卡尺	90	30
20	铣加工	以 $\phi250$ 外圆装夹,校准 $\phi21$ 孔中心;铣上表面,保零件总厚度为 20mm;铣削 $R107$、$\phi35$,钻 $2\times\phi10$ 孔	铣	加工中心游标卡尺	90	90
25	清理	清理工件,锐角倒钝	钳			
30	检验	检验工件尺寸	检			

本训练任务针对第 20 工序铣加工,进行工序设计,制订工序卡片,如表 4-5 所示。

表 4-5 铣加工工序卡片

机械加工 工序卡片	产品型号	CLJG-01	产品型号	CLJG-01	第 1 页
	产品名称	槽轮机构	零部件名称	驱动轮	共 1 页

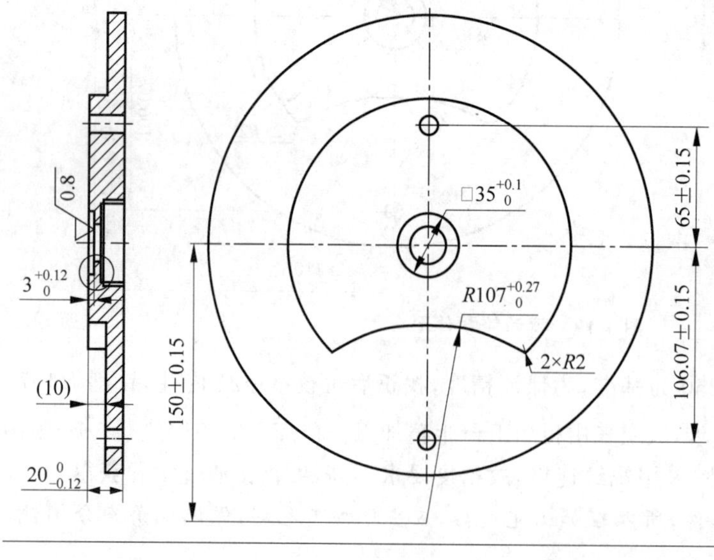

工序号	20	
工序名	铣加工	
材料	C45	
设备	加工中心	
设备型号	VAL650e	
夹具	三爪自定心卡盘	
量具	游标卡尺	
准结工时	90min	
单件工时	90min	

续表

工步	工步内容	刀具	$S/$ (r/min)	$F/$ (mm/r)	$a_p/$ mm	$a_e/$ mm	工步工时/min	
							机动	辅助
1	工件安装							5
2	铣削 ϕ160 上表面，厚度 20mm 留 0.1mm	ϕ63 面铣刀	800	200	2	30	15	
3	粗铣 R107，底面留 0.1mm，侧面留 0.2mm	ϕ63 面铣刀	800	200	2	30	20	
4	精铣 ϕ160 表面	ϕ63 面铣刀	1500	200	0.1	30	5	
5	精铣 R107 底面，侧面留 0.2mm	ϕ63 面铣刀	1500	200	0.1	30	5	
6	粗铣 ϕ35，侧面、底面留 0.1mm	ϕ12 立铣刀	2000	500	0.5	5	10	
7	精铣 ϕ35 侧面	ϕ12 立铣刀	2000	200	0	0.1	5	
8	精铣 ϕ35 底面	ϕ12 立铣刀	2000	200	0.1	0	5	
9	精铣 R107 侧面	ϕ12 立铣刀	2000	200	0	0.1	5	
10	加工 ϕ10 中心孔	中心钻	1000	100	0.05	2	2	
11	加工 ϕ10 孔	ϕ10 麻花钻	1000	200	0.1	6	3	
12	R2 倒圆							5
13	拆卸、清理工件							5

4.10 数控加工程序编写

根据工序加工工艺，编写加工程序，如表 4-6 所示。

表 4-6 驱动轮数控加工程序

序号	程 序 语 句	注　　解
	O0001;	
	T01 M6;	调用 ϕ63 肩铣刀
	G10 L13 P01 R0.2;	设定侧面余量为 0.2mm
	S800 M3;	设置主轴正转，转速为 800r/min
	G0 G90 G54 X115 Y0;	定位到点(X115,Y0)处
	G43 Z10 H01;	在长度补偿号 01 条件下，刀具运动到点(X115,Y0,Z10)处位置
N1	G1 Z0.1 F200;	粗加工顶面
	X80;	第一次进刀
	G2 I−80 J0;	切削一周
	G1 X50;	第二次进刀
	G2 I−50 J0;	切削一周

续表

序号	程序语句	注解
	G1 X20；	第三次进刀
	G2 I-20 J0；	切削一周
	G1 X0；	第四次进刀
	Z10；	
N2	G0 X0 Y-115；	粗加工 $R107$
	G1 Z-2；	加工第一层
	G1 G41 X79.738 Y-78.65 D1；	
	G3 X-79.738 R107；	
	G1 G40 X0 Y-115；	
	G1 Z-4；	加工第二层
	G1 G41 X79.738 Y-78.65 D1；	
	G3 X-79.738 R107；	
	G1 G40 X0 Y-115；	
	G1 Z-6；	加工第三层
	G1 G41 X79.738 Y-78.65 D1；	
	G3 X-79.738 R107；	
	G1 G40 X0 Y-115；	
	G1 Z-8；	加工第四层
	G1 G41 X79.738 Y-78.65 D1；	
	G3 X-79.738 R107；	
	G1 G40 X0 Y-115；	
	G1 Z-9.9；	加工第五层，留 0.1mm 底面余量
	G1 G41 X79.738 Y-78.65 D1；	
	G3 X-79.738 R107；	
	G1 G40 X0 Y-115；	
	Z10；	
N3	S1500 M3；	精加工 $\phi160$ 表面
	G0 X115 Y0；	
	G1 Z0 F200；	
	X80；	
	G2 I-80 J0；	
	G1 X50；	
	G2 I-50 J0；	
	G1 X20；	
	G2 I-20 J0；	
	G1 X0；	
	Z10；	
N4	G10 L13 P01 R0.1；	$R107$ 侧面半精加工
	G0 X0 Y-115；	
	G1 Z-10；	
	G1 G41 X79.738 Y-78.65 D1；	
	G3 X-79.738 R107；	

续表

序号	程序语句	注　解
	G1 G40 X0 Y-115；	
	Z10；	
	M5；	
	T02 M6；	调用第二把刀，φ12立铣刀
	G10 L13 P02 R0.2；	设定半径补偿磨耗为0.2mm，即侧面余量
	S2000 M3；	
	G0 G90 G54 X0 Y0；	
	G43 Z10 H2；	
N5	G1 Z-2 F500；	粗加工φ35
	G1 G41 X7.5 D2；	
	G3 I-7.5 J0；	
	G1 G40 X0 Y0；	
	G1 Z-2.9 F500；	
	G1 G41 X7.5 D2；	
	G3 I-7.5 J0；	
	G1 G40 X0 Y0；	
	G1 Z10；	
N7	G10 L13 P02 R0；	精加工φ35侧面
	G1 Z-2.9 F500；	
	G1 G41 X7.5 D2；	
	G3 I-7.5 J0；	
	G1 G40 X0 Y0；	
N8	G1 Z-3 F500；	精加工φ35底面
	G1 G41 X7.5 D2；	
	G3 I-7.5 J0；	
	G1 G40 X0 Y0；	
N9	G0 X0 Y-115；	精加工R107侧面
	G1 Z-10；	
	G1 G41 X79.738 Y-78.65 D1；	
	G3 X-79.738 R107；	
	G1 G40 X0 Y-115；	
	G0 Z10；	
	M5；	
N10	T3 M6；	打中心孔
	S1200 M3；	
	G0 G90 G54 X0 Y65；	
	G43 Z10 H3；	
	G81 Z-1.5 R1 F100；	
	Y-106.07 Z-11.5 R-9；	
	G80；	
	M5；	
N11	T4 M6；	打中心孔

续表

序号	程序语句	注　解
	S1200 M3；	
	G0 G90 G54 X0 Y65；	
	G43 Z10 H4；	
	G83 Z-23 R1 Q5 F200；	
	Y-106.07 R-9；	
	G80；	
	M5；	
	G0 Z100；	
	M30；	

任务三　上 机 训 练

4.11　设备与用具

设备：AVL650e 立式加工中心。

刀具：ϕ63 肩铣刀、ϕ12 立铣刀、中心钻、ϕ10 麻花钻。

夹具：K11-320 三爪自定心卡盘（配反爪）。

工具：什锦锉刀。

量具：0～150mm 游标卡尺、0.02mm 杠杆百分表（配安装杆）。

毛坯：ϕ250×23（加工后）。

辅助用品：卡盘扳手、橡胶锤、毛刷等。

4.12　认识机床

4.12.1　开机检查

检查机床外观各部位（如防护罩、脚踏板等部位）是否存在异常；检查机床润滑油、冷却液是否充足；检查刀架、夹具、导轨护板上是否有异物；检查机床面板各旋钮状态是否正常；在开机后，检查机床是否存在报警等。可参考表 4-7 对机床状态进行点检。

表 4-7　机床开机准备卡片

检查项目		检查结果	异常描述
机械部分	主轴部分		
	进给部分		
	换刀机构		
	夹具系统		
电器部分	主电源		
	冷却风扇		

续表

检 查 项 目		检 查 结 果	异 常 描 述
数控系统	电气元件		
	控制部分		
	驱动部分		
辅助部分	冷却系统		
	压缩空气		
	润滑系统		

在机床开机前点检正常后,可通过旋转机床后面电气开关旋钮 ，打开机床电源。

按下机床控制面板上的电源开按键 ，使数控系统上电。

按照按钮标识方向旋转紧急停止开关旋钮 ，解决紧急状态。

开机后系统面板如图 4-20 所示。

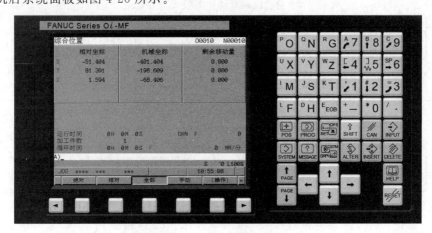

图 4-20　启动后的系统面板

4.12.2　机床面板

在正式操作机床前,应熟悉加工中心操作面板各按键的功能及操作方法,熟记紧急按键的具体位置。

1. 操作模式选择开关旋钮

图 4-21 所示为机床的操作模式选择开关旋钮,旋钮标识指向哪个模式,在显示屏会有相应标识。

(1)"连线"模式(RMT):用于在线加工或调用 CF 卡中程序加工。

(2)"编辑"模式(EDIT):用于编辑程序或外部数控读入。

(3)"记忆"模式(MEM):用于自动运行读入内存的程序。

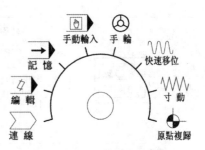

图 4-21　操作模式选择旋钮

(4)"手动输入"模式(MDI)：用于运行控制器 MDI 面板录入的程序。

(5)"手轮"模式(HAND)：用于使用电子手轮运动机床轴。

(6)"快速移位"模式(JOG)：使用操作面板方向按键快速移动机床坐标轴。

(7)"寸动"模式(INC)：使用操作面板按给定的距离点动机床坐标轴。

(8)"原点复归"模式(REF)：用于机床返回参考点。

2. 其他功能按键

操作面板其他常用功能按键如表 4-8 所示，由于机床控制系统、生产厂家、出厂批次的不同，因此，操作面板功能按键的位置、数量会有一些差异。在操作机床前，应熟悉当前机床的操作说明书，并按安全操作要求使用机床。

表 4-8 操作面板其他常用功能按键

图　示	说　明
MST锁住	名称："MST 锁住"功能按键 本功能仅在自动相关模式下有效。 (1)当功能按键指示灯亮时，"MST 锁住"功能按键有效。在此功能打开后，当程序中编有 M、S、T 辅助代码时，程序中的 M 代码、S 代码、T 代码指令将被忽略而不执行，其余指令可以正常执行。 (2)当功能按键指示灯不亮时，"MST 锁住"功能按键无效。在此功能关闭后，程序中的 M 代码、S 代码、T 代码指令正常执行
選擇跳過	名称："选择跳过"功能按键 本功能仅在自动相关模式下有效。 (1)当功能按键指示灯亮时，"选择跳过"功能按键有效。在此功能打开后，在自动运行中，当程序段的开头指定了一个"/"(斜线)符号时，此程序段将略过，不被执行。 (2)当功能按键指示灯不亮时，"选择跳过"功能按键无效。在此功能关闭后，当程序单节前有"/"(斜线)符号时，此程序段也可以正常执行
單節執行	名称："单节执行"功能按键 本功能仅在自动相关模式下有效。 (1)当功能按键指示灯亮时，"单节执行"功能按键有效。在此功能打开后，程序将按单节执行，执行完当前单节后程序暂停，继续按程序启动功能按键后方可执行下一单节程序，以后执行程序以此类推。 (2)当功能按键指示灯不亮时，"单节执行"功能按键无效。加工程序将一直执行到程序结束。
選擇停	名称："选择停"功能按键 本功能仅在自动模式有效。 (1)当功能按键指示灯亮时，"选择停"功能按键有效。在此功能打开后，在执行程序中，若有 M01 指令，则程序将停止于该单节；若想要继续执行程序，则按下程序启动功能按键即可。 (2)当功能按键指示灯不亮时，"选择停"功能按键无效。在此功能关闭后，即使程序中有 M01 指令，程序也不会停止执行

续表

图　示	说　明
Z轴锁定	名称："Z轴锁定"功能按键 （1）当功能按键指示灯亮时，"Z轴锁定"功能按键有效。在此功能打开后，此时，Z轴锁定不能移动，但在显示器画面中，Z轴的相对坐标/绝对坐标会随着指令或手动移动而发生实时变化。 （2）在解除此功能后，需要重新回归机械零点，在回零点正确且完毕后，再进行其他相关操作。 （3）如果未回零点，而进行了相关操作，则会造成坐标偏移，甚至出现撞机、程序乱跑等异常现象，从而导致危险。 （4）当功能按键指示灯不亮时，"Z轴锁定"功能按键无效
機械鎖定	名称："机械锁定"功能按键 （1）当功能按键指示灯亮时，所有轴"机械锁定"功能按键有效。在此功能打开后，无论在手动模式或自动模式中移动任意一个轴，CNC均停止向该轴伺服电机输出脉冲（移动指令），但依然在进行指令分配，对应轴的绝对坐标和相对坐标也相应得到更新。 （2）M、S、T、B代码会继续执行，不受"机械锁定"的限制。 （3）在解除此功能后，需要重新回归机械零点，在回零点正确且完毕后，再进行其他相关操作。 （4）如果未回零点，而进行了相关操作，则会造成坐标偏移，甚至出现撞机、程序乱跑等异常现象，从而导致危险
空運行	名称："空运行"功能按键 本功能仅在自动相关模式下有效。 （1）当功能按键指示灯亮时，"空运行"功能按键有效。在此功能打开后，程序中所设定的F值（切削进给率）指令无效，其各轴移动速率依慢速位移速率所指定速率位移。 （2）在功能有效时，若程序执行循环程序，则慢速进给或切削进给率无法改变实际进给率，依然按照控制中的F值以固定进给率位移。
主軸正轉 主軸停止 主軸反轉	名称："主轴正转"功能按键、"主轴停止"功能按键、"主轴反转"功能按键 （1）在机床执行一次S代码后，选中"手动输入"模式，按下"主轴正转"/"主轴反转"功能按键后，主轴进行顺时针/逆时针旋转。主轴旋转速度＝先前执行的主轴速度S值×主轴修调旋钮所在的挡位。 （2）主轴无论处于正转或反转状态，按下"主轴停止"功能按键均可以停止正在旋转中的主轴。 （3）使用条件如下。 ① 本功能按键仅在"手动输入"模式、"快速移位"模式、"寸动"模式下才能使用。 ② 本功能按键在自动操作时无效
自動冷卻 手動冷卻	名称："手动冷却"功能按键 在"手动输入""快速移位""寸动"模式下，按下此功能按键，指示灯亮，冷却液打开，此时，冷却自动按键指示灯闪烁。 名称："自动冷却"功能按键 在自动模式下，按下此功能按键指示灯亮，冷却液自动操作有效。当程序执行到M08或按下冷却功能按键时，冷却液会自动喷出；当程序执行到M09时冷却液会自动关闭

续表

图　　示	说　　明
![−X / +X 按键]	名称：＋／－X 控制功能按键 在 JOG 模式下，若按住此功能按键，则 X 轴依进给倍率/快速倍率的速度向机床 X 轴"＋／－"方向（正/负方向）移动，同时，本功能按键指示灯点亮；当松开按键后，X 轴停止向机床 X 轴"＋／－"方向移动，同时，本功能按键指示灯熄灭。 另外，当程序执行 X 轴方向移动程序指令时，该功能按键指示灯也将点亮；当停止移动指令时，该功能按键指示灯熄灭。 此外，"＋X"按键也作为 X 轴回零点触发键
![−Y / +Y 按键]	名称：＋／－Y 控制功能按键 在 JOG 模式下，若按住此功能按键，则 Y 轴依进给倍率/快速倍率的速度向机床 Y 轴"＋／－"方向（正/负方向）移动，同时，本功能按键指示灯点亮；当松开按键后，Y 轴停止向机床 Y 轴"＋／－"方向移动，同时，本功能按键指示灯熄灭。 另外，当程序执行 Y 轴方向移动程序指令时，该功能按键指示灯也将点亮；当停止移动指令时，该功能按键指示灯熄灭。 此外，"＋Y"功能按键也作为 Y 轴回零点触发键
![−Z / +Z 按键]	名称：＋／－Z 控制功能按键 在 JOG 模式下，若按住此功能按键，则 Z 轴依进给倍率/快速倍率的速度向机床 Z 轴"＋／－"方向（正/负方向）移动，同时，本功能按键指示灯点亮；当松开按键后，Z 轴停止向机床 Z 轴"＋／－"方向移动，同时，本功能按键指示灯熄灭。 另外，当程序执行 Z 轴方向移动程序指令时，该功能按键指示灯也将点亮；当停止移动指令时，该功能按键指示灯熄灭。 此外，"＋Z"功能按键也作为 Z 轴回零点触发键
![O.T.REL. 按键]	名称：超程释放功能按键 (1) 当本机床各轴的行程超过硬体极限时，机床会出现超程报警，同时，此功能按键的指示灯熄灭，机床动作停止。这时，按住此功能按键在"手轮"模式下用手持单元将机床超程的轴往反方向移动。 (2) 在本机床上电后，松开紧急停止开关旋钮，机床正常开启无报警，该功能按键指示灯常亮
![进给倍率旋钮]	名称：进给倍率及进给修调旋钮 (1) 此旋钮位于本机床操作面板上，控制编程指定 G01 指令速度，实际进给速度＝编程给定 F 代码值×进给倍率开关所在倍率值％。 (2) 在"寸动"模式下，此时控制 JOG 进给倍率，实际 JOG 进给速度＝参数设定固定值×进给倍率开关所在倍率值％。 (3) 配合轴进给控制功能按键使用
![快速倍率旋钮]	名称：快速倍率旋钮 (1) 此旋钮位于本机床操作面板上，控制编程指定 G00 指令速度，实际进给速度＝参数设置 G00 指令最大速度值×快速进给倍率开关所在倍率值％。 (2) 在快送模式下，此时控制手动快速进给倍率，实际快速进给速度＝参数设置 G00 指令最大速度值×快速进给倍率开关所在倍率值％。快移动倍率可以在 F0、25％、50％、100％四个挡位进行调整。 (3) 配合轴进给控制功能按键使用

续表

图 示	说 明
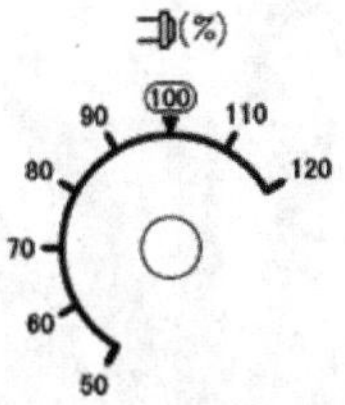	名称：主轴修调旋钮 (1) 此旋钮位于本机床操作面板上，控制编程制定的主轴转速 S，实际转速＝编程给定 S 代码值×主轴修调开关所在的倍率值％ (2) 当编程设定速度超过主轴最高转速时，即转速达到 100％以上倍率时，主轴修调速度等于主轴最高转速。 (3) 配合主轴控制按键中的按键使用

4.12.3 回原点

在机床启动后，应手动使各轴回原点。先将操作模式选择开关旋钮转至"原点复归"模式，然后按操作面板"＋Z"功能按键令 Z 轴回原点，再按"＋X""＋Y"功能按键令 X、Y 轴回原点。

4.13 刀具准备

在加工前，应先将本任务所需刀具准备齐全。表 4-9 所示为所需要的刀具列表，按表 4-9 中序列，正确安装刀具。

表 4-9 刀具列表

刀 号	刀 柄	拉 钉	刀 具	安 装 工 具
T1				
T2				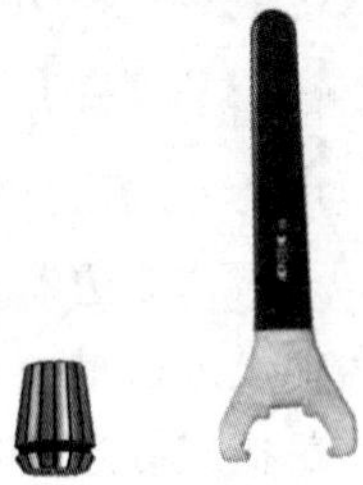

续表

刀 号	刀 柄	拉 钉	刀 具	安装工具
T3				
T4				

刀具应在工作台使用正确的工具及方法安装,错误操作可能损坏刀具、刀柄,甚至造成人员伤害,刀具的安装精度对加工精度及刀具使用寿命也有较大的影响。

4.14 寻边器与工件找正

4.14.1 安装寻边器

使用 ER 夹头,将机械式寻边器安装在 BT 刀柄上,如图 4-22 所示,将装好的刀柄放入加工中心 24 号刀位备用。

机械式寻边器是一种高精度测量工具,能快速且容易地设定机械主轴与加工物基准面精确的中心位置。机械式寻边器的结构包括夹持部与偏心部的中空上、下检测头;可贯穿上、下检测头的弹簧;可分别钩固定住弹簧且嵌置在上、下检测头顶、底端的顶盖、底盖等构件。机械式寻边器的工作原理:可夹持在

图 4-22 机械式寻边器

机床上低速旋转,自动通过偏心作用调整,以找正加工中心位置。

4.14.2 加工中心调用寻边器

首先将加工中心操作模式切换到 MDI 手动输入挡,在系统面板 MDI 程序输入位置中,输入程序段"T24 M6;",按循环启动功能按键运行当前程序段,机床会执行从刀库中调用安装有寻边器的刀柄的操作。

再输入程序段"S500 M3;",按循环启动功能按键运行当前程序段,使寻边器以 500r/min 的速度低速旋转。

4.14.3 工件找正

在"快速移位"模式下,点动 X、Y、Z 方向控制功能按键,使旋转的寻边器移动到工件的左侧附近,切换到"手轮"模式,用手轮缓慢移动 X 轴,使寻边器贴近工件,直到寻边器上、下两部分同心,如图 4-23(a)所示。

按下系统面板上的 按键,单击屏幕底部的 按钮,显示机床相对坐标;输入 后单击 按钮,此时,相对坐标中的 X 坐标值变为 0。

只通过移动 X 轴和 Z 轴,使寻边器在工件右侧达到上、下同心位置,如图 4-23(b)所示,记录此时相对坐标中 X 的坐标值,此处为 259.816(见图 4-24),将其数值除以 2 后为 129.908。

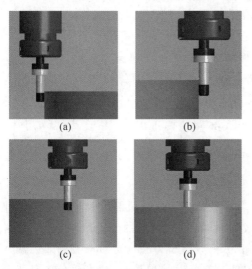

图 4-23 寻边器找正

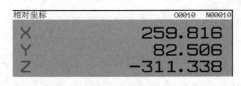

图 4-24 X 轴相对坐标

按下系统面板上 按键,单击屏幕底部 按钮,在 G54 坐标系中 X 轴位置输入"X129.908"后,单击 按钮,完成 X 方向的工件找正。

用类似的方法,将寻边器 Y 方向移动到靠近操作者一侧,X 方向相对坐标值显示"X129.908"处,用手轮缓慢移动 Y 轴方向,使寻边器上、下同心,如图 4-23(c)所示,将此时的相对坐标 Y 值预置为 0。然后通过只移动 Y 轴和 Z 轴,使寻边器在远离工件一侧与工件接触,达到上、下同心状态,如图 4-23(d)所示。记下此时的相对坐标 Y 轴显示的数值,并将其数值的 1/2 输入到工件坐标系 Y 轴位置,单击"测量"按钮,系统将自动计算工件坐标系 G54 的偏置值。

在 Z 轴方向向上移动,使寻边器完全离开工件。

为验证工件找正结果的正确性,可以在 MDI 模式下运行程序段"G0 G90 G54 X0 Y0;",此时,工件坐标系(绝对坐标)中 X 轴、Y 轴显示的数据应当为 0,刀具中心与工件中心重合,如图 4-25 所示。

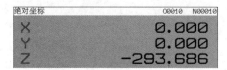

图 4-25 工件找正后验证

4.15 对刀

以对第一把刀为例,介绍加工中心的对刀方法。

在对刀前,应将工件表面清理干净,保证无毛刺和切屑,并且,要校准 Z 轴设定仪的零点位置,常用的 Z 轴设定仪的标定高度为 50mm。将 Z 轴设定仪放在工件上的对刀位置,保证底面与工件紧密贴合。

在对刀时,首先在 MDI"手动输入"模式下输入程序段"T1 M6;"调出第一把刀,使用"快速移位"模式使刀具快速接近 Z 轴设定仪,在离上方有一定距离时,采用"手轮"模式,向下移动 Z 轴,使刀齿向下压 Z 轴设定仪活动块。如图 4-26 所示,观察示数,当指针归零时,停止向下移动,此时,刀齿与工件上表面的距离为 50mm。

此时,按下系统面板上 按键,单击显示屏底部 按钮,光标移动到刀偏列表第一行的"形状(H)"位置,在数据输入框

图 4-26 对刀示意图

中输入"Z10",单击显示屏底部的"测量"按钮,完成第一把刀的对刀。部分加工中心刀具长度补偿设置没有"测量"功能,这时只需记录当前的绝对坐标 Z 轴的数值,然后减去 50mm,将结果直接输入到"形状(H)"对应的位置即可。

注意:对于多齿且尺寸较大的铣刀,在对刀时,应当以最低刀齿的对刀数据为准。输入长度补偿数据后,还应在"形状(D)"列输出对应刀具的半径补偿数值,以便在进行刀具半径补偿时使用。

用相同的方法,可以完成 T2~T4 的对刀操作,完成的对刀结果,如图 4-27 所示。

刀偏号	形状(H)	磨耗(H)	形状(D)	磨耗(D)
001	−270.013	0.000	31.500	0.000
002	−293.125	0.000	6.000	0.000
003	−254.300	0.000	1.500	0.000
004	−233.540	0.000	5.000	0.000
005	0.000	0.000	0.000	0.000

图 4-27 对刀刀偏数据

4.16 程序编辑

在机床系统面板上进行程序的录入共有两种操作,一种是程序号的录入,另一种是程序语句的录入。

将操作模式选择开关旋钮旋至"编辑"模式,在系统面板上按下 [PROG] 按键,在显示屏底部 [EDIT 程序 目录] 可点击"目录"按钮切换程序存放的路径,通过单击"程序"按钮进行程序的编辑。

在录入程序时,首先单击"程序"按钮进入程序的编辑界面。

先输入程序号 [O*0*0*0*1] 后,直接按下 [INSERT] 按键,再按下 [EOB] 按键和 [INSERT] 按键实现换行。此时要注意,程序号(名)是分两步输入的,如果输入"O0001;",则按下 [INSERT] 按键会提示"格式错误" [A)O0001; 格式错误]。

程序体其他语句的输入与程序名不同,在直接输入整段语句后,直接按下 [INSERT] 按键即可,如 [K T ;1 I M SP 6 E EOB INSERT]。

完成输入的程序如图 4-28 所示。

图 4-28 输入完成的数控程序

如在编辑过程中需要删除程序块,则将光标移动到要删除的程序块位置(黄色高亮显示),然后按下 [DELETE] 按键完成删除;如需修改光标选中的程序块,则在 [A)___] 处输入想要更改后的内容,再按下 [ALTER] 按键,即可完成程序块内容的替换;如需删除待输入区 [A)X100Y3000_] 中的最后一个字符,则按下 [CAN] 按键,即可实现退格。

4.17 数控加工程序的仿真

FANUC-0i-MF 系统提供了数控加工程序的图形校验功能,由于系统版本和机床生产商出厂配置的不同,因此,图形校验界面也有差异。

在加工中心操作面板上,选择"记忆"模式(MEM),即程序的自动执行模式,调入需要进行图形校验的数控加工程序,然后在系统面板上按下 [CSTM/GRAPH] 按键,进入图形校验界面,如图 4-29 所示,可通过图形校验界面参数设备仿真区域大小和视角。

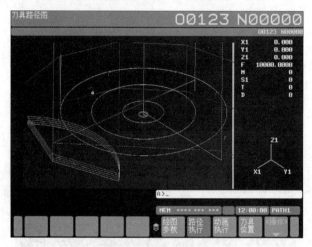

图 4-29 图形校验界面

在操作面板上按下 [机械锁定] 按键和 [空运行] 按键,开启程序的坐标轴锁定状态和空运行状态,将进给倍率调节到 0%,按下程序循环启动功能按键,程序即开始图形校验,通过调节进给倍率旋钮调整图形校验速度。

4.18 零件加工

在图形校验过程验证无问题后,即可进行零件加工。在零件加工前,应详细了解机床的安全操作要求,穿戴好劳动保护服装和用具。在进行零件加工时,应熟悉加工中心各操作按键的功能和位置,了解紧急状况的处置方法。

注意:如已进行图形校验操作,则在操作完成后,须执行机床回归机械零点操作,然后再进行其他的相关操作。如果未回零点,而进行了相关操作,则会造成坐标偏移,甚至出现撞机、程序乱跑等异常现象,从而导致危险。

在操作面板上选择"记忆"模式(MEM),即程序的自动执行模式,调入需要进行加工的数控加工程序,按循环启动功能按键进行自动加工。

数控加工程序的首次自动运行应在调试状态下进行。机床进给倍率旋钮先调低至 0%,按下操作面板上 [单段执行] 功能按键,通过单击显示屏底部"程序检查"按钮切换到自动运行状态检查界面,如图 4-30 所示。

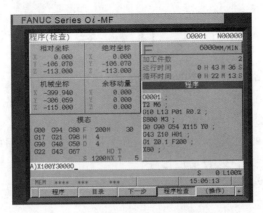

图 4-30　程序自动运行状态检查界面

在此状态下,每按一次循环启动功能按键,程序都只自动执行光标所在的一行。在按下循环启动功能按键前,应观察刀具与工件间的距离是否安全;在按下循环启动功能按键后,通过进给倍率旋钮控制机床的运动速度,同时,对照显示屏"余移动量"栏显示的剩余移动量,观察刀具与工件之间的实际距离。若实际距离与剩余移动量相差过大,则应果断停机检查,以免发生撞机事故。在程序调试过程中,还应密切注意显示屏的"模态"状态显示,确保主轴转速、进给速度、工件坐标系号、补偿状态及补偿号等无异常发生。

4.19　零件检测

在零件加工完成后,应当认真清理工件,并按照质量管理的相关要求,对加工完成的零件进行相关检验,保证生产质量。机械加工零件"三级"检验卡片如表 4-10 所示。

表 4-10　机械加工零件"三级"检验卡片

零部件图号		零部件名称		工　序　号	
材料		送检日期		工序名称	
检验项目	自检结果	互检结果	专业检验		备注
检验结论	□合格　　□不合格　　□返修　　□让步接收　　　　　　　　　　　检验签章:　　　　　　　　　　　　　　　　　年　　　月　　　日				
不符合项描述					

项目总结

通过驱动轮数控铣削加工，需要掌握数控铣削程序的基本格式和基本切削加工指令的使用方法。能够应用 G01 指令、G02 指令、G03 指令基本切削加工指令进行零件切削；能够应用 G81 指令、G83 指令进行简单孔的加工。

掌握立式加工中心的基本操作方法，包括开关机、刀具安装、工件找正、对刀、程序编辑、图形校验、数控加工程序调试及自动运行等。

通过任务训练，养成良好的职业素养，培养正确的加工中心安全操作规范，养成基本的机械加工质量意识。

课后习题

1. 填空题

(1) 加工中心与一般的铣床最大的区别在于加工中心可以_____。

(2) 加工中心能完成的加工工序有_____、_____、_____、_____等。

(3) 立式加工中心与相对应的卧式加工中心相比，_____，_____，_____。

(4) 加工中心的自动换刀装置由驱动机构、_____组成。

(5) 按刀具的用途及加工方法分类，成型车刀属于_____类型。

2. 判断题

(1) 立式加工中心立柱高度有限，相应地，对箱体类工件加工范围比较小，这是立式加工中心的缺点。（　　）

(2) 换刀点是加工中心手动换刀的地方。（　　）

(3) 为了使机床达到热平衡状态，必须使机床运转 3min 以上。（　　）

(4) 在试切和加工中，在刃磨刀具和更换刀具后，一定要重新测量刀具长度，并修改刀具相关补值。（　　）

(5) 利用一般计算工具，运用各种数学方法，人工进行刀具轨迹的运算，并进行指令编程称为机械编程。（　　）

3. 选择题

(1) 当前，加工中心进给系统的驱动方式多采用（　　）。
  A. 液压伺服进给系统　　　　B. 电气伺服进给系统
  C. 气动伺服进给系统　　　　D. 液压电气联合式

(2) 按主轴的种类分类，加工中心可分为单轴、双轴、（　　）加工中心。
  A. 不可换主轴箱　　　　　　B. 三轴、五面
  C. 复合、四轴　　　　　　　D. 三轴、可换主轴箱

(3) （　　）指令表示直线插补的指令。
  A. G9　　　B. G111　　　C. G01　　　D. G93

(4)(　　)指令表示换刀的指令。
　　A. G50　　　　　B. M06　　　　　C. G66　　　　　D. M62
(5)在机床通电后,应首先检查(　　)是否正常。
　　A. 加工路线　　　　　　　　　　B. 各开关按钮和键
　　C. 电压、油压、加工路线　　　　D. 工件精度

4. 简答题

(1)简述数控铣削的加工特点。

(2)简述数控加工中心夹具。

(3)简述加工循环指令 G81 指令和 G83 指令的区别。

自我学习检测评分表如表 4-11 所示。

表 4-11　自我学习检测评分表

项　目	目 标 要 求	分值	评分细则	得分	备注
学习关键知识点	(1)了解 AVL650e 立式加工中心的结构及主要参数 (2)理解铣削加工的特点 (3)熟悉常用铣刀的分类并能进行刀具的正确选择 (4)了解数控铣床的通用夹具及量具 (5)掌握数控铣削程序基本格式及基本准备功能指令、辅助功能指令的使用	25	理解与掌握		
工艺准备	(1)能够正确识读基本平面类零件图 (2)能够独立确定加工工艺过程,并正确填写工艺文件 (3)能够根据工序加工工艺,编写正确的加工程序	25	理解与掌握		
上机训练	(1)会正确选择相应的设备与用具 (2)掌握铣削刀具的选择和安装方法 (3)能够根据零件结构特点和精度合理选用量具,并正确、规范地测量出相关尺寸 (4)能够正确操作数控加工中心,并根据加工情况调整加工参数	50	(1)理解与掌握 (2)操作流程		

思政小课堂

项目五　槽轮铣削编程与加工训练

> **思维导图**

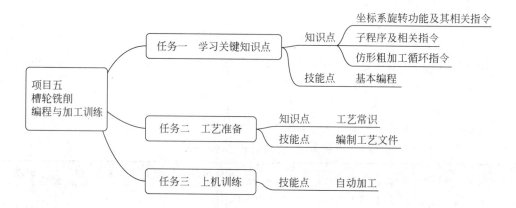

> **学习目标**

知识目标

（1）了解槽轮的工作原理。
（2）了解槽轮在本机构中的用途和关键要求。

能力目标

（1）掌握坐标系旋转指令的使用方法。
（2）掌握子程序指令的使用方法。
（3）掌握铣削刀具的选择与使用方法。
（4）能够独立确定加工工艺路线，并正确填写工艺文件。
（5）能够正确操作数控加工中心，并根据加工情况调整加工参数。
（6）能够根据零件结构特点和精度合理选用量具，并正确、规范地测量相差尺寸。

素养目标

（1）培养学生的科学探究精神和态度。
（2）培养学生的工程意识。
（3）培养学生的团队合作能力。

任务一 学习关键知识点

5.1 数控机床坐标系旋转功能及原理

在数控编程时,为了描述机床的运动,简化程序编制的方法及保证记录数据的互换性,数控机床的坐标系和运动方向均已标准化,ISO 和我国都拟定了命名的标准。机床坐标系(machine coordinate system,MCS)是以机床原点 O 为坐标系原点,并遵循右手笛卡儿直角坐标系建立的,由 X 轴、Y 轴、Z 轴组成的直角坐标系。机床坐标系是用来确定工件坐标系的基本坐标系,是机床上固有的坐标系,并设有固定的坐标原点。

在实际应用中,往往会遇到一些零件属于相同几何尺寸的特征,且分布在某一点周围的情况,也有些零件的工程图采用了旋转的局部视图来表示几何尺寸。这时,为了编程的方便,常常采用将原有工件坐标系旋转的方法进行数控编程。

对于立式加工中心,常用的旋转变换为 XOY 平面内的变换。如图 5-1 所示,为点 $A(x,y)$ 绕点 $O'(x_0,y_0)$ 旋转 θ 角到 $A'(x',y')$ 的示意图。

假设旋转的中心位于坐标原点,坐标 (x,y) 旋转到 (x',y') 后,可以很容易等到旋转后的坐标为

$$\begin{cases} x' = x\cos\theta - y\sin\theta \\ y' = x\sin\theta + y\cos\theta \end{cases}$$

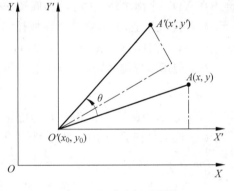

图 5-1 机床坐标系旋转

这个变换为基本的旋转变换,写成矩阵形式为

$$\begin{bmatrix} x' & y' \end{bmatrix} = \begin{bmatrix} x & y \end{bmatrix} \begin{bmatrix} \cos\theta & \sin\theta \\ -\sin\theta & \cos\theta \end{bmatrix}$$

在实际应用中,回转的中心经常不是工作坐标系的原点,如图 5-1 所示,旋转中心在点 $O'(x_0,y_0)$ 的位置。所以,在进行坐标系旋转时,需要进行坐标的转换,即临时将坐标系移动到旋转中心 $O'(x_0,y_0)$ 处,再按旋转中心位于坐标原点时的方法进行旋转。此时,向数控系统输送坐标,再将坐标系数值转换为原工件坐标系中的坐标值。

将坐标中心移动到临时位置 $O'(x_0,y_0)$ 后,原点 A 的坐标值可表示为

$$\begin{cases} x' = x - x_0 \\ y' = y - y_0 \end{cases}$$

这个变换写成矩阵形式为

$$[x'\ y'\ 1] = [x\ y\ 1] \begin{bmatrix} 1 & 0 & 0 \\ 0 & 1 & 0 \\ -x_0 & -y_0 & 1 \end{bmatrix}$$

所以,在 XOY 工件坐标系中以几何方法表示出旋转后的点 $A'(x',y')$ 坐标为

$$\begin{cases} x' = (x-x_0)\cos\theta - (y-y_0)\sin\theta + x_0 \\ y' = (x-x_0)\sin\theta + (y-y_0)\cos\theta + y_0 \end{cases}$$

用矩阵的形式可以表示为

$$[x'\ y'\ 1] = [x\ y\ 1] \begin{bmatrix} 1 & 0 & 0 \\ 0 & 1 & 0 \\ -x_0 & -y_0 & 1 \end{bmatrix} \begin{bmatrix} \cos\theta & \sin\theta & 0 \\ -\sin\theta & \cos\theta & 0 \\ 0 & 0 & 1 \end{bmatrix} \begin{bmatrix} 1 & 0 & 0 \\ 0 & 1 & 0 \\ x_0 & y_0 & 1 \end{bmatrix}$$

则将

$$\boldsymbol{T} = \begin{bmatrix} 1 & 0 & 0 \\ 0 & 1 & 0 \\ -x_0 & -y_0 & 1 \end{bmatrix} \begin{bmatrix} \cos\theta & \sin\theta & 0 \\ -\sin\theta & \cos\theta & 0 \\ 0 & 0 & 1 \end{bmatrix} \begin{bmatrix} 1 & 0 & 0 \\ 0 & 1 & 0 \\ x_0 & y_0 & 1 \end{bmatrix}$$

称为在 XOY 平面中绕点 (x_0,y_0) 旋转的变换矩阵。

在数控机床上,通常将这类变换计算过程写入模态调用宏程序中,并将其定义为准备功能代码 G,FANUC 0i 系统所对应的 G 代码为 G68 指令、G69 指令。指令格式如下。

(1) 应用坐标旋转功能,指令格式为:

G68 X<x_0> Y<y_0> R<θ>;

其中,(x_0,y_0) 为旋转中心坐标,θ 为旋转角度,以逆时针方向为正方向。

(2) 取消坐标旋转功能,指令格式为:

G69;

如图 5-2 所示,在加工直角 $\triangle ABC$ 时,按传统编程方法需要使用三角函数方法计算出点 B、点 C 的坐标值,然后进行切削。在应用坐标系旋转方法编程时,可以将 $\triangle ABC$ 看作其全等 $\triangle AB'C'$ 绕点 A 逆时针旋转 30°得到的。此时,点 B'、点 C' 的坐标值可以非常方便地计算出来,其程序如表 5-1 所示。

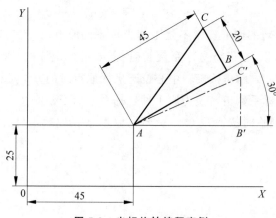

图 5-2 坐标旋转编程实例

表 5-1　坐标系旋转指令编程实例

程 序 语 句	注　　解
G0 X45 Y25;	快速定位到起点 A 处
G1 Z-2 F500;	Z 向切入工件
G68 X45 Y25 R20;	定义坐标系旋转模式启动,以点(45,25)为基准旋转 20°
G1 X90;	向点 B′ 切削
Y45;	向点 C′ 切削
G69;	取消坐标系旋转模式
G1 X45 Y25;	向点 A 切削

5.2　子程序的格式与应用

在一个加工程序中,如果其中有些加工内容完全相同或相似,为了简化程序,可以把这些重复的程序段单独列出,按一定的格式编写成单独的程序,并单独加以命名,这组程序段称为子程序。主程序在执行过程中如果需要某一子程序,则可以通过调用指令来调用该子程序。子程序执行完后,又返回主程序,继续执行后面的程序段。通常子程序不可以作为独立的加工程序使用,它只能通过调用,实现加工中的局部动作。

5.2.1　子程序的格式

子程序和主程序在格式上并无本质区别,主要是结束标记不同。主程序使用 M02 指令或 M30 指令表示程序结束,而子程序使则用 M99 指令表示程序结束,并自动返回调用它的程序。如表 5-2 所示为一个主程序调用一个子程序后,两者之间的格式对比。

表 5-2　主程序与子程序格式对比

主　程　序	子　程　序
O0001;	O1234;
T1 M6;	G1 Z F2 F500;
S2000 M3;	G1 X80 F1000;
G0 G90 G54 X20 Y20;	Y80;
G43 Z10 H1;	X20;
M98 P0011234;	Y20;
G0 Z100;	G1 Z10;
M5;	M99;
M30;	

在命名子程序时,应避开系统指定的特征功能,或者机床生产商个性化定义的子程序号。

5.2.2　子程序的调用

FANUC 0i 系统子程序常用的调用指令格式为:

M98 P○○○△△△△;

其中,○○○为调用次数,数值范围为001～999,在输入时数字前边的"0"可以省略,当调用次数为1次时,"1"也可以省略;△△△△为子程序号,取值范围为0001～9999,在输入时数字前的"0"不可以省略。

当调用次数更多时,采用参数 L 指定调用次数,子程序调用格式为:

M98 P△△△△L○○○○○○○○;

当程序号也大于5位时,子程序调用的格式为:

M98P△△△△△△△L○○○○○○○○;

当程序不以数字命名,而使用字符型命名时,子程序调用格式为:

M98 <△△△△> L○○○○○○○○;

子程序不仅可以被主程序调用,同时,也可以被其他子程序调用,这种方法称为子程序的嵌套,如图5-3所示。根据系统的不同,其子程序的嵌套级数也不相同,一般 FANUC 0 系统可以嵌套4级,较新的系统如 FANUC 0i MF 可以嵌套15级。

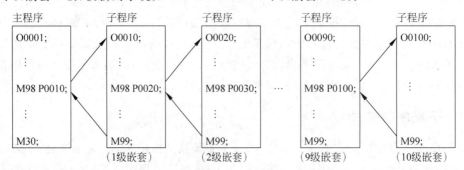

图 5-3　FANUC 0i MF 子程序嵌套示意图

5.2.3　子程序在相似形状重复加工时的应用

在相似形状重复加工时,多涉及分布在以不同的起点开始的位置;所以,通常在主程序中使用绝对坐标指令编程进行定位;在子程序中使用相对坐标指令编程,实现形状重复。加工如图5-4所示的零件,通常可参考表5-3的方法使用子程序编程。

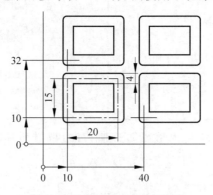

图 5-4　相似形状重复加工编程

表 5-3 相似形状重复加工编程示例

主 程 序		子 程 序	
O0001；	主程序号	O1234；	子程序号
T1 M6；	调用 T1 刀具	G1 Z-2 F500；	Z 向切入工件
S2000 M3；	主轴以 2000r/min,正转	G1 G91 X20 F1000；	以相对坐标方向向右移动 20mm
G0 G90 G54 X10 Y10；	定位到 G54 坐标系绝对坐标(10,10)	Y15；	向上移动 15mm
G43 Z10 H1；	长度补偿号 H01,到起始位置	X-20；	向左移动 20mm
M98 P1234；	调用子程序 O1234 一次	Y-15；	向下移动 15mm
G0 X40 Y10；	快速定位到点(40,10)处	G1 G90 Z10；	以绝对坐标抬刀至起始位置
M98 P1234；	调用子程序 O1234 一次	M99；	子程序返回
G0 X40 Y32；	快速定位到点(40,32)处		
M98 P1234；	调用子程序 O1234 一次		
G0 X10 Y32；	快速定位到点(10,32)处		
M98 P1234；	调用子程序 O1234 一次		
G0 Z100；	快速抬刀到安全位置		
M5；	主轴停转		
M30；	主程序结束,复位		

5.2.4 子程序在分层加工时的应用

如图 5-5 所示零件,在加工时,刀具无法一次性切削至 Z 轴方向 30mm 深度处,若以每层切深 2mm 编写程序,则相同的零件面 XY 轮廓需要重复 15 次。如果零件轮廓复杂,则会造成程序行数量过大。此时,使用子程序功能可以将轮廓加工的部分全部放在子程序中,在子程序中第一行以相对坐标指令指定每层切削的深度,然后以绝对坐标指令编写轮廓加工程序。在子程序结束时,不要编写抬刀指令,而且在子程序结束时,终点的坐标(x,y)要与子程序开始时相同。

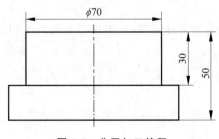

图 5-5 分层加工编程

在主程序中使用 M98 指令调用子程序,指定调用次数。应注意调用次数必须为整数,因此,需要灵活控制每层切深,选择的每层切深可以整除总切深。

$$\text{int}[调用次数]=总切削深度÷每层切削深度$$

以子程序方式编写的分层加工示例如表 5-4 所示。

表 5-4 分层加工编程示例

主 程 序		子 程 序	
O0001;		O1234;	
T1 M6;		G1 G91 Z-2 F500;	每层相对当前位置下切深 2mm
S2000 M3;		G1 G90 G41 X45 Y10 D1;	绝对坐标指令方式,引入刀补
G0 G90 G54 X60 Y0;	选择毛坯外点(60,0)处下刀	G3 X35 Y0 R10;	以 R10 圆弧切入轮廓
G43 Z10 H1;		G2 I-35 J0;	切削轮廓
G1 Z0 F500;	提前运动到分层加工起始位置	G3 X45 Y-10 R10;	以 R10 圆弧切出轮廓
M98 P151234;	调用子程序 O1234,共调用 15 次	G1 G40 X60 Y0;	取消刀补,返回起始点
G1 Z10;		M99;	子程序返回
G0 Z100;			
M5;	抬刀		
M30;			

任务二 工 艺 准 备

5.3 零件图分析

如图 5-6 所示,根据零件的使用要求,选择 45 钢作为槽轮零件的毛坯材料,毛坯下料尺寸定为 $\phi 215 \times 25$。

零件 $\phi 35$ 为关键尺寸,需要与轴承配合,可以在车削加工中通过镗孔获得;零件外圆尺寸在整个机床中为大间隙配合,所以粗加工即可;4 处 R80 为间隙配合,通过粗铣获得;20 槽要求位置精度较高,与拨销滑动摩擦配合,所以,要求有较低的侧面表面粗糙度,需要进行精铣加工。

注意:此零件为手动操作件,所以在加工完成后,应去除加工毛刺,保证锐角充分倒钝,以确保在使用过程中的人身安全。

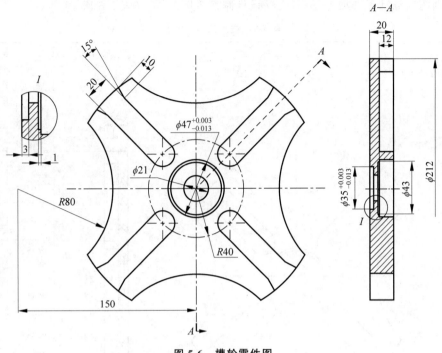

图 5-6 槽轮零件图

5.4 工艺设计

根据零件图分析,确定工艺过程,如表 5-5 所示。

表 5-5 工艺过程卡片

机械加工 工艺过程卡片	产品型号		零部件序号		第 页	
	产品名称		零部件名称		共 页	
材料 牌号		毛坯规格		毛坯质量	数量	
工序号	工序名	工序内容		工段	工艺装备	工时/min	
						准结	单件

本训练任务针对铣加工,进行工序设计制订工序卡片,如表 5-6 所示。

表 5-6 铣加工工序卡片

机械加工 工序卡片	产品型号		零部件序号		第 页
	产品名称		零部件名称		共 页

工步	工步内容	刀具	S/ (r/min)	F/ (mm/r)	a_p/ mm	a_e/ mm	工步工时/min	
							机动	辅助

右侧栏目：工序号、工序名、材料、设备、设备型号、夹具、量具、准结工时、单件工时

5.5 数控加工程序编写

根据工序加工工艺,编写加工程序,如表 5-7 所示。

表 5-7 槽轮数控加工程序

段号	程序语句	注　释
	O0001;	主程序
	T1 M6;	
	S1500 M3;	
	G0 G90 G54 X120 Y0;	
	G43 Z10 H1;	
N1	G1 Z0 F500;	定位至 $R80$ 圆弧铣削起点
	M98 P111001;	调用加工 $R80$ 圆弧,共调用 11 次,加工到深度 $Z-22$
	G1 Z10;	
	G68 X0 Y0 R90;	坐标旋转 90°加工第二个
	G1 Z0 F500;	
	M98 P111001;	
	G1 Z10;	
	G68 X0 Y0 R180;	坐标旋转 180°加工第三个
	G1 Z0 F500;	
	M98 P111001;	
	G1 Z10;	
	G68 X0 Y0 R270;	坐标旋转 270°加工第四个
	G1 Z0 F500;	
	M98 P111001;	
	G1 Z10;	
	G69;	
N2	G68 X0 Y0 R45;	坐标旋转 45°,粗加工槽
	G0 X120 Y0;	
	G1 Z0 F500;	
	M98 P111002;	调用槽加工程序,共调用 11 次,每次切削 2mm
	G1 Z10;	
	G68 X0 Y0 R135;	
	G0 X120 Y0;	
	G1 Z0 F500;	
	M98 P111002;	
	G1 Z10;	
	G68 X0 Y0 R225;	
	G0 X120 Y0;	
	G1 Z0 F500;	
	M98 P111002;	
	G1 Z10;	
	G68 X0 Y0 R315;	
	G0 X120 Y0;	
	G1 Z0 F500;	
	M98 P111002;	
	G1 Z10;	
	G69;	
N3	G68 X0 Y0 R45;	
	G0 X120 Y0;	

续表

段号	程序语句	注　　释
	M98 P1003;	精加工45°槽
	G68 X0 Y0 R135;	
	G0 X120 Y0;	
	M98 P1003;	精加工135°槽
	G68 X0 Y0 R225;	
	G0 X120 Y0;	
	M98 P1003;	精加工225°槽
	G68 X0 Y0 R315;	
	G0 X120 Y0;	
	M98 P1003;	精加工315°槽
	G0 Z100;	
	M5;	
	M30;	
	%	
	O1001;	粗加工R80圆弧子程序
	G1 G91 Z-2 F500;	
	G1 G90 G41 X100 Y62.45 D1;	
	G3 Y-62.45 R80;	
	G1 G40 X120 Y0;	
	M99;	
	%	
	O1002;	粗加工槽子程序
	G1 G91 Z-2 F500;	
	G1 X40;	
	X120;	
	M99;	
	%	
	O1003;	精加工槽子程序
	G1 Z-21 F500;	
	G1 G41 X120 Y16.557 D1;	
	X95.527 Y10;	
	X40;	
	G3 Y-10 R10;	
	G1 X95.527;	
	X120 Y-16.557;	
	G1 G40 X120 Y0;	
	G1 Z10;	
	M99;	

任务三 上 机 训 练

5.6 设备与用具

设备：AVL650e 立式加工中心。
刀具：ϕ16 立铣刀。
夹具：K11-320 三爪自定心卡盘（配芯轴）。
工具：什锦锉刀。
量具：0～150mm 游标卡尺、0.02mm 杠杆百分表（配安装杆）。
毛坯：ϕ212×20（车加工后）。
辅助用品：卡盘扳手、橡胶锤、毛刷等。

5.7 开机检查

检查机床外观各部位（如防护罩、脚踏板等部位）是否存在异常；检查机床润滑油、冷却液是否充足；检查刀架、夹具、导轨护板上是否有异物；检查机床面板各旋钮状态是否正常；在开机后,检查机床是否存在报警等。可参考表 5-8 对机床状态进行点检。

表 5-8 机床开机准备卡片

检查项目		检查结果	异常描述
机械部分	主轴部分		
	进给部分		
	换刀机构		
	夹具系统		
电器部分	主电源		
	冷却风扇		
数控系统	电气元件		
	控制部分		
	驱动部分		
辅助部分	冷却系统		
	压缩空气		
	润滑系统		

5.8 零件加工

输入数控加工程序,运行程序完成零件的数控加工。

5.9 零件检测

在零件加工完成后,应当认真清理工件,并按照质量管理的相关要求,对加工完成的零件进行相关检验,保证生产质量。机械加工零件"三级"检验卡片如表5-9所示。

表5-9 机械加工零件"三级"检验卡片

零部件图号		零部件名称		工 序 号	
材料		送检日期		工序名称	
检验项目	自检结果	互检结果	专业检验	备注	
检验结论	□合格　　□不合格　　□返修　　□让步接收 　　　　　　　　　　　　　　检验签章: 　　　　　　　　　　　　　　　　　年　　月　　日				
不符合项描述					

项目总结

通过槽轮数控铣削加工,需要掌握数控铣削程序的基本格式和基本切削加工指令的使用方法。能够熟练应用基本切削加工指令 G1 指令、G2 指令、G3 指令进行零件切削;能够灵活应用坐标旋转指令 G68 指令、G69 指令及子程序指令 M98 指令、M99 指令编写加工程序。

掌握立式加工中心的基本操作方法,包括开关机、刀具安装、工件找正、对刀、程序编辑、图形校验、数控加工程序调试及自动运行等。

通过任务训练,养成良好的职业素养,培养正确的加工中心安全操作规范,养成基本的机械加工质量意识。

课后习题

1. 填空题

(1) 在机床通电后,应首先检查_____是否正常。

(2) 为应对采用了旋转局部视图来表示几何尺寸的工程图,常常采用将原有工件_____的方法进行数控编程。

(3) 子程序和主程序在格式并无本质区别,主要是_____不同,主程序使用_____指令表示程序结束,而子程序使用 M99 指令表示程序结束,并自动返回调用它的程序。

(4) 系统子程序常用的调用指令格式为：_____。

(5) 在开机检查时,应该检查机床_____(如防护罩、脚踏板等部位)是否存在异常;检查机床_____、_____是否充足;检查刀架、夹具、导轨护板上是否有异物;检查机床面板各旋钮状态是否正常;在开机后,检查机床是否存在报警等。

2．判断题

(1) 子程序返回主程序的指令为 M98 指令。　　　　　　　　　　　　　(　　)

(2) 在加工完成后,应去除加工毛刺,保证锐角充分倒钝,以确保在使用过程中的人身安全。　　　　　　　　　　　　　　　　　　　　　　　　　　　　(　　)

(3) 通常子程序可以作为独立的加工程序使用。　　　　　　　　　　　(　　)

(4) 子程序不仅可以被主程序调用,同时,也可以被其他子程序调用,这种方法称为子程序的嵌套。　　　　　　　　　　　　　　　　　　　　　　　　　(　　)

(5) 使用 M98 指令调用子程序,指定调用次数,调用次数必须为整数。　(　　)

3．选择题

(1) 机床坐标系的原点称为(　　)。

  A. 工件零点　　　B. 编程零点　　　C. 机械原点　　　D. 空间零点

(2) 取消坐标系旋转指令为(　　)指令。

  A. G65　　　　　B. G66　　　　　C. G68　　　　　D. G69

(3) 子程序返回指令为(　　)指令。

  A. M96　　　　　B. M97　　　　　C. M98　　　　　D. M99

(4) 在用轨迹法切削槽类零件时,槽两侧表面,(　　)。

  A. 一面为顺铣、一面为逆铣　　　B. 两面均为顺铣

  C. 两面均为逆铣　　　　　　　　D. 不需要做任何加工

(5) 加工内廓类零件时,(　　)。

  A. 不用留有精加工余量

  B. 为保证顺铣,刀具要沿工件表面左右摆动

  C. 为保证顺铣,刀具要沿内廓表面逆时针运动

  D. 刀具要沿工件表面任意方向移动

4．简答题

简述子程序的应用场合。

自我学习检测评分表如表 5-10 所示。

表 5-10 自我学习检测评分表

项　目	目 标 要 求	分值	评 分 细 则	得分	备注
学习关键知识点	(1) 掌握坐标系旋转指令的使用方法 (2) 掌握子程序指令的使用方法	20	理解与掌握		
工艺准备	(1) 能够正确识读零件图 (2) 能够独立确定加工工艺过程，并正确填写工艺文件 (3) 能够根据工序加工工艺，编写正确的加工程序	30	理解与掌握		
上机训练	(1) 能够根据零件结构特点和精度合理选用量具，并正确、规范地测量相关尺寸 (2) 掌握槽轮铣削的操作流程 (3) 能够正确操作数控加工中心，并根据加工情况调整加工参数	50	(1) 理解与掌握 (2) 操作流程		

思政小课堂

项目六　驱动轴、从动轴螺纹孔编程与加工训练

> **思维导图**

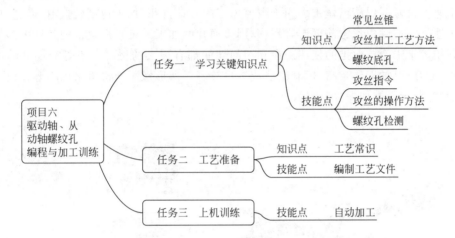

> **学习目标**

知识目标

(1) 了解驱动轴、从动轴零件的工作要求。

(2) 了解驱动轴、从动轴在本机构中的用途和关键要求。

能力目标

(1) 掌握孔加工指令的使用方法。

(2) 掌握螺纹加工指令的使用方法。

(3) 掌握螺纹加工刀具的选择与使用方法。

(4) 能够独立确定加工工艺路线,并正确填写工艺文件。

(5) 能够正确操作数控加工中心,并根据加工情况调整加工参数。

(6) 能够根据零件结构特点和精度合理选用量具,并正确、规范地测量相差尺寸。

素养目标

(1) 培养学生的科学探究精神和态度。

(2) 培养学生的工程意识。

(3) 培养学生的团队合作能力。

任务一 学习关键知识点

6.1 常见丝锥

丝锥是一种加工内螺纹的工具,按照形状可分为螺旋槽丝锥、刃倾角丝锥、直槽丝锥和管用螺纹丝锥等;按照使用环境可分为手工丝锥和机用丝锥;按照规格可分为公制丝锥、美制丝锥和英制丝锥等。丝锥是制造业操作者在攻丝时采用的最主流的加工工具。

直槽丝锥加工容易、精度略低、产量较大,一般用于普通车床、钻床及攻丝机的螺纹加工,切削速度较慢。螺旋槽丝锥多用于数控加工中心钻盲孔,加工速度较快、精度高、排屑较好、对中性好。刃倾角丝锥前部有容屑槽,用于通孔的加工。工具厂提供的丝锥大多为涂层丝锥,其相较于未涂层丝锥的使用寿命和切削性能都有很大的提高。不等径设计的丝锥切削负荷分配合理、加工质量高,但其制造成本也较高。梯形螺纹丝锥常采用不等径设计。直槽丝锥和螺旋槽丝锥如图 6-1 所示。

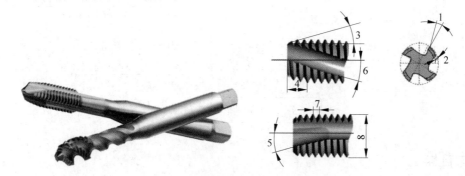

图 6-1 直槽丝锥和螺旋槽丝锥

除了丝锥的公称直径(图 6-1 中"8"所示)和螺距(图 6-1 中"7"所示),在选用丝锥时,还要分清手工丝锥和机用丝锥。分辨手工丝锥和机用丝锥可通过其倒角长度(图 6-1 中"4"所示),一般手工丝锥倒角较长,机用丝锥倒角较短。另外,还要注意其公差,国际标准丝锥常用的公差带号为 ISO1、ISO2、ISO3,ISO1 用于加工 4H、5H 螺纹孔,ISO2 用于加工 5G、6H 螺纹孔,ISO3 用于加工 6G、7H、7G 螺纹孔。有时还可以看到一些未标注公差带的丝锥,有的资料将其定义为 ISO4,用于手工丝锥,加工 6H、7H 螺纹孔。

6.2 攻丝加工工艺方法

图 6-2 所示为攻丝加工常见的几种工艺方法。

(1)螺类丝锥加工通孔:这是一种非常强劲的攻丝工艺,适用于比较苛刻的加工工况,在攻丝过程的同时扩大底孔,将切屑穿过孔向前推出。

(2)螺旋槽丝锥加工盲孔:这是一种常用的机械攻丝工艺,使用螺旋槽丝锥配合机床刚性攻丝功能,使切屑比较轻巧地沿丝锥螺旋槽向上排出。

(3)通用直槽攻丝加工:这种工艺方向可用于手工攻丝和机械攻丝,可以加工所有孔型,特别适用于短切屑材料(如铸铁),经常用于汽车行业和泵、阀生产行业。切削碎屑在重力作用下向下落,所以在自上而下攻盲孔时,要注意清屑。

(4) 挤压攻丝:这是一种无切屑的攻丝工艺,主要用于低碳钢、不锈钢、铝等塑性材料,可以用于各种类型的孔和孔深。在加工铝合金时,由于形变强化的作用,可以增加其螺纹的强度。挤压攻丝所用丝锥称为成型丝锥,与其对应的其他丝锥称为切削丝锥。

(a) 螺类丝锥加工通孔　　(b) 螺旋槽丝锥加工盲孔　　(c) 通用直槽攻丝加工　　(d) 挤压攻丝

图 6-2　常见攻丝加工工艺方法

6.3　攻丝加工螺纹底孔尺寸

在使用攻丝方法攻盲孔螺纹时,螺纹底孔深度和底孔直径尺寸的计算公式分别为

$$H_{深} = h_{有效} + 0.7D$$

$$D_{底孔} = D - P$$

式中,$h_{有效}$ 为螺纹的有效深度;D 为螺纹的公称直径;P 为螺纹的螺距。

在实际应用中,螺纹盲孔的底孔可加工得略深一些,一方面防止攻丝时丝锥尖顶顶死,另一方面可以增加底部容屑空间。对于不同材料,在加工螺纹底孔时,孔径也往往略有修正:一般在加工铸铁、青铜等脆性材料时,计算孔径需减去 1.1 倍的螺距;在加工钢、铜等塑性材料时,计算孔径需减去 1 倍的螺距。所以这些参数要根据生产实际,结合材料及其热处理状态灵活选择。

6.4　螺纹攻丝的操作方法

6.4.1　手工攻丝

手工攻丝的操作如图 6-3 所示。

(1) 将丝锥装入丝锥铰手上。操作要点:丝锥铰手用于夹住丝锥便于转动丝锥攻削。丝锥铰手要夹在锥柄的方榫上,不要夹在光滑的锥柄上;否则,在攻丝时,丝锥铰手与丝锥之间会打滑。

(2) 将夹在丝锥铰手上的丝锥垂直地插入底孔。操作要点:用目测法从纵、横两个

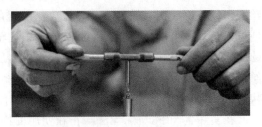

图 6-3　手工攻丝操作

方向交叉检查丝锥与孔口平面的垂直程度。如不垂直,则应予以纠正。

(3) 双手靠拢握住丝锥铰手柄,大拇指抵住丝锥铰手中部向下施压,按顺时针方向,边转边压,使丝锥逐步切入孔内。操作要点:以均等的压力集中在丝锥铰手中部,力求使丝锥垂直地切入孔内。压力要适当大些,转动丝锥铰手要缓慢,防止孔口滑牙。

(4) 丝锥切入孔内 1~2 牙,检查丝锥的垂直程度。若发现偏斜,则应予以纠正。操作要点:用目测法交叉检查丝锥垂直程度。如果刀齿切入过多,强行纠正,则会损坏丝锥。纠正方法:边转动丝锥铰手,边朝偏斜的反方向缓缓地纠正。

(5) 在丝锥攻入孔内 3~4 牙后,双手分开握住丝锥铰手柄,不再加压,均匀地转动铰手。每转动 3/4 圈,倒旋 1/4 圈,攻削至旋不动为止。在攻削过程中,要适量地加入润滑油。操作要点:在攻入孔口 3~4 牙后,已有部分螺纹形成,此时,只需转动丝锥铰手,不要加压,丝锥会自行向下切入。若此时仍再加压攻削,会损坏已形成的螺纹。在攻削时有切屑形成,会卡阻丝锥,倒旋的目的是切断切屑,减少阻力。加入润滑油,减少切削阻力,可减少螺纹粗糙度,延长丝锥使用寿命。

(6) 双手扶持丝锥铰手柄,按逆时针方向均匀、平稳地转动,从孔内退出丝锥,清理丝锥和螺孔内切屑。操作要点:双手要均匀平稳地倒旋丝锥铰手。当丝锥即将从孔内完全退出时,应避免丝锥晃动,损坏螺纹。

6.4.2 机械攻丝

攻丝是数控加工中心上常见的孔加工内容,把选定的丝锥安装在专用攻丝刀套上前,要确认浮动刀套具有拉伸和压缩等特性。攻丝一般步骤如下。

第 1 步:X、Y 定位。

第 2 步:选择主轴转速和旋转方向。

第 3 步:快速移动至点 R 处。

第 4 步:进给运动至指定深度。

第 5 步:主轴停止。

第 6 步:主轴反向旋转。

第 7 步:进给运动返回。

第 8 步:主轴停止。

第 9 步:快速返回至初始位置。

第 10 步:重新开始主轴正常旋转。

机械攻丝分柔性攻丝和刚性攻丝。柔性攻丝多用于开环或半闭环系统,相对于刚性攻丝的闭环系统成本要低很多。柔性攻丝由于存在累积误差、主轴转速较低,所以加工效率较低,适用于中、小批量生产;刚性攻丝由于主轴转速与进给速度之间建立严格的线性关系,选用优质合金刀具可以实现高转速、高进给,所以加工效率高,适用于大批量生产。柔性攻丝涉及的机床参数较少,对调试人员要求不高。另外,在默认状态下,系统一般为柔性攻丝状态;而刚性攻丝涉及的机床参数较多,相对于柔性攻丝较为烦琐。

在实际生产中,机床无法与正在使用的特定丝锥节距精确匹配。在机床所加工的螺纹与丝锥实际节距之间总存在细微的差异。如果采用整体丝锥夹,则该差异对丝锥寿命及螺纹质量具有决定性的影响,这是因为在丝锥上要施加额外的轴向作用力;如果采用带张力压缩浮动的丝锥夹,则丝锥寿命及螺纹质量将大大提高,这是因为消除了丝锥上这些额外的轴向作用力。带传统张力压缩浮动的丝锥夹存在的问题是,它们会引起攻丝深度方面较大的变化。随着丝锥变钝,将丝锥启动到孔内所需要的压力增加,在丝锥开始切削之前,在丝锥驱动器内所用的压缩行程将会更大,其结果会导致攻丝深度较浅。

刚性攻丝的主要优点之一是,在盲孔加工中,可以精确控制深度。为了精确而一致地加工工件,需要采用具有足够补偿的丝锥夹来实现较高的丝锥寿命,且不在深度控制方面引起任何变化。因此,在本章训练任务中,我们选择刚性攻丝进行加工。

6.5 攻丝指令

攻丝指令 G84 指令和 G74 指令可以在标准方式和刚性方式下进行攻丝加工,G84 指令用于加工右旋螺纹,G74 指令用于加工左旋螺纹。在执行攻丝指令后,加工中心主轴的旋转或停止受攻丝指令控制,不必人为干预。

G84 指令与 G74 指令除加工的螺纹方向不同之外,其他参数完全相同,在此以 G84 指令为例进行介绍,其指令格式为:

G84 X__ Y__ Z__ R__ P__ F__ K__;

其中,X__ Y__ 为螺纹孔位置;Z__ 为从点 R 到孔底的距离或孔底位置;R__ 为攻丝开始位置,开始位置应在正式主攻丝前面;P__ 为孔底及返回点 R 时的暂停时间;F__ 为切削进给速度;K__ 为重复次数(仅限需要重复时)。

其动作过程如图 6-4 所示。

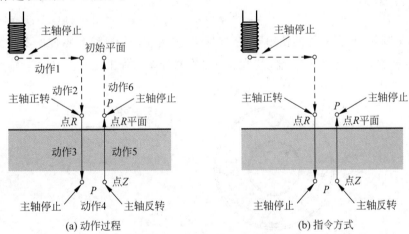

图 6-4 G84 攻丝指令动作

默认方式为柔性攻丝方式,如需要进行刚性攻丝,则可通过以下 3 种方式进行指定。

(1) 在攻丝指令之前,指定"M29 S__;"

(2) 在含有攻丝指令的程序段中,指定"M29 S__;"

(3) 在系统参数中,将指令指定为刚性攻丝,如将参数 G84(No.5200♯0)置 1,则使用 G84 指令默认即为刚性攻丝方式。

注意:在进入攻丝加工指令后,进给倍率调节是无效的。

6.6 螺纹孔的检测

通常螺纹孔使用螺纹塞规进行测量。如图 6-5 所示,螺纹塞规是测量内螺纹尺寸正确性的工具。螺纹塞规可分为普通粗牙螺纹塞规、细牙螺纹塞规和管子螺纹塞规 3 种。螺距为 0.35mm 及以下、2 级精度及以上的螺纹塞规,螺距为 0.8mm 及以下的 3 级精度的螺纹

图 6-5 螺纹塞规

塞规都没有止端测头。长度为 100mm 以下的螺纹塞规为锥柄螺纹塞规,长度为 100mm 以上的为双柄螺纹塞规。螺纹塞规可检测被测螺纹的最大实体牙型,以及其作用中径是否超过其最大实体牙型的中径。同时,通过螺纹塞规的"通规"和"止规",还可检测被测螺纹的底径实际尺寸是否超过其最大实体尺寸。

使用螺纹塞规不可暴力操作,在通常情况下,螺纹塞规的通规可在被测螺纹的任意位置转动,若通过全部螺纹长度,则判定为合格,否则为不合格;在螺纹塞规的止规与被测螺纹对正后,若旋入螺纹长度在两个螺距之间止住,则为合格,不可强行用力通过,否则判为不合格。

任务二 工艺准备

6.7 零件图分析

如图 6-6 所示,根据零件的使用要求,选择 45 钢为驱动轴螺纹孔零件的毛坯材料,毛坯尺寸为 $\phi 50$ 棒料,在螺纹孔加工前,其他部位已在数控车床加工完成。

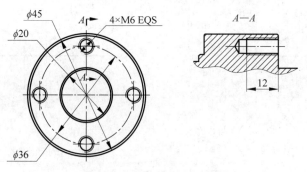

图 6-6 驱动轴螺纹孔零件图

本工序螺纹孔为紧固连接孔,整体加工精度不高,该训练任务主要练习加工中心攻丝基本操作,因此,相对较简单。

注意:此零件为手动操作件,所以在加工完成后,应去除加工毛刺,保证锐角充分倒钝,以确保在使用过程中的人身安全。

6.8 工艺设计

根据零件图分析,确定工艺过程,如表6-1所示。

表6-1 铣加工工序卡片

机械加工工序卡片	产品型号	CLJG-01	零部件序号	AXIS-01	第1页
	产品名称	槽轮机构	零部件名称	驱动轴	共1页
			工序号		
			工序名	孔加工	
			材料	C45	
			设备	立式加工中心	
			设备型号	VAL650e	
			夹具	三爪自定心卡盘	
			量具	螺纹塞规	
			准结工时	60min	
			单件工时	10min	

工步	工步内容	刀具	S/(r/min)	F/(mm/r)	a_p/mm	a_e/mm	工步工时/min 机动	辅助
1	工件安装							1
2	加工中心孔	中心钻	1500	100	0.05	1	1	
3	钻 $\phi 5$ 底孔	$\phi 5$ 麻花钻	1200	100	0.1	2.5	3	
4	倒角	$\phi 10 \times 90°$	1200	100	0.1	2	1	
5	刚性攻丝	M6	600	60	1	0.8	2	
6	拆卸,清理工件							2

6.9 数控加工程序编写

根据工序加工工艺,编写加工程序,如表6-2所示。驱动轴与从动轴只是在对刀时有所不同,两者的数控加工程序与工序文件相同。

表6-2 驱动轴、从动轴螺纹孔数控加工程序

段号	程序语句	注释
	O001;	主程序
N1	T1 M6;	调用中心钻
	S1500 M3;	
	G0 G90 G54 X18 Y0;	
	G43 Z10 H1;	
	G81 Z-1.5 R1 F100;	

续表

段号	程序语句	注　释
	X0 Y18;	
	X-18 Y0;	
	X0 Y-18;	
	G80;	
	G0 Z100;	
	M5;	
N2	T2 M6;	调用 $\phi 5$ 麻花钻
	S1200 M3;	
	G0 G90 G54 X18 Y0;	
	G43 Z10 H2;	
	G83 Z-17.7 Q2 R1 F100;	使用深孔钻削指令进行底孔加工,深度为 17.7mm(尖点)
	X0 Y18;	
	X-18 Y0;	
	X0 Y-18;	
	G80;	
	G0 Z100;	
	M5;	
N3	T3 M6;	调用 $\phi 10 \times 90°$ 倒角钻
	S1000 M3;	
	G0 G90 G54 X18 Y0;	
	G43 Z10 H3;	
	G81 Z-3.5 R1 F100;	
	X0 Y18;	
	X-18 Y0;	
	X0 Y-18;	
	G80;	
	G0 Z100;	
	M5;	
N4	T4 M6;	调用丝锥
	S600 M3;	
	G0 G90 G54 X18 Y0;	
	G43 Z10 H4;	
	M29 S600;	调用刚性攻丝指令
	G84 Z-16 R1 F60;	攻丝深度在设置时,要综合考虑丝锥的尖顶,保证有效螺纹为 10mm
	X0 Y18;	
	X-18 Y0;	
	X0 Y-18;	
	G80;	
	G0 Z100;	
	M5;	

续表

段号	程序语句	注　释
	M30;	

任务三　上机训练

6.10　设备与用具

设备：AVL650e 立式加工中心。

刀具：$\phi3$ 中心钻、$\phi5$ 麻花钻、$\phi10\times90°$ 倒角钻、M6 机用丝锥。

夹具：K11-320 三爪自定心卡盘。

工具：什锦锉刀。

量具：0～150mm 游标卡尺、0.02mm 杠杆百分表(配安装竿)、M6-6H 螺纹塞规。

毛坯：$\phi45$(车加工后)。

辅助用品：卡盘扳手、橡胶锤、毛刷等。

6.11　开机检查

检查机床外观各部位(如防护罩、脚踏板等部位)是否存在异常；检查机床润滑油、冷却液是否充足；检查刀架、夹具、导轨护板上是否有异物；检查机床面板各旋钮状态是否正常；在开机后，检查机床是否存在报警等。可参考表 6-3 对机床状态进行点检。

表 6-3　机床开机准备卡片

	检查项目	检查结果	异常描述
机械部分	主轴部分		
	进给部分		
	换刀机构		
	夹具系统		
电器部分	主电源		
	冷却风扇		
数控系统	电气元件		
	控制部分		
	驱动部分		
辅助部分	冷却系统		
	压缩空气		
	润滑系统		

6.12 零件加工

输入数控加工程序,运行程序完成零件的数控加工。

6.13 零件检测

在零件加工完成后,应当认真清理工件,并按照质量管理的相关要求,对加工完成的零件进行相关检验,保证生产质量。机械加工零件"三级"检验卡片如表6-4所示。

表6-4 机械加工零件"三级"检验卡片

零部件图号		零部件名称		工 序 号	
材料		送检日期		工序名称	
检验项目	自检结果	互检结果	专业检验	备注	
检验结论	□合格　□不合格　□返修　□让步接收 检验签章: 　　　　　　　　　　　　年　　月　　日				
不符合项描述					

项目总结

通过驱动轴、从动轴螺纹孔加工,需要掌握加工中心螺纹加工程序的基本格式和基本切削加工指令的使用方法。能够熟练应用螺纹切削加工指令 G84 指令、G74 指令进行螺纹孔攻丝;能够灵活应用 M29 指令编写刚性攻丝加工程序。

掌握立式加工中心的基本操作方法,包括开关机、刀具安装、工件找正、对刀、程序编辑、图形校验、数控加工程序调试及自动运行等。

通过任务训练,养成良好的职业素养,培养正确的加工中心安全操作规范,养成基本的机械加工质量意识。

课 后 习 题

1. 填空题

(1) 常见的攻丝加工工艺方法有_____、_____、_____、_____。

(2) 在螺纹底孔深度的计算公式中，h 表示_____，D 表示_____，P 表示_____。

(3) 对于不同材料，在加工螺纹底孔时，孔径也往往略有修正：一般在加工铸铁、青铜等脆性材料时，计算孔径需减去_____倍的螺距；在加工钢、铜等塑性材料时，计算孔径需减去_____倍的螺距。

(4) 螺纹攻丝的操作方法有：手工攻丝、_____。

(5) 攻丝指令 G84 指令中"K"的含义是_____。

2. 判断题

(1) 直槽丝锥加工容易、精度略低、产量较大，一般用于普通车床、钻床及攻丝机的螺纹加工用，切削速度较慢。( )

(2) 通用直槽攻丝加工，切削碎屑在重力作用下向下落，所以在自上而下攻盲孔时要注意清屑。( )

(3) 在手工攻丝时，加入润滑油后均匀地转动丝锥绞手，丝锥会自行向下切入，只需要攻削至旋不动为止。( )

(4) 在实际生产中，机床无法与正在使用的特定丝锥节距精确匹配。在机床所加工的螺纹与丝锥实际节距之间总存在细微的差异。( )

(5) G84 指令与 G74 指令除加工的螺纹方向不同，即 G84 指令左旋、G74 指令右旋以外，其他参数完全相同。( )

3. 选择题

(1) 进行孔类零件加工时，钻孔→平底钻扩孔→倒角→精镗孔的方法适用于( )。
    A. 阶梯孔   B. 小孔径的盲孔
    C. 大孔径的盲孔   D. 较大孔径的平底孔

(2) 机械攻丝分柔性攻丝和刚性攻丝，下列中错误的是( )。
    A. 柔性攻丝多用于开环或半闭环系统，相对于刚性攻丝的闭环系统成本要低很多
    B. 柔性攻丝由于存在累积误差、主轴转速较低，所以加工效率较低，适用于中小批量生产
    C. 刚性攻丝由于主轴转速与进给速度之间建立严格的线性关系，选用优质合金刀具可以实现高转速、高进给，所以加工效率高，适用于大批量生产
    D. 在默认状态下，系统一般为刚性攻丝状态

(3) 攻丝指令 G84 指令动作在 G98 指令方式下可划分为( )部分。
    A. 5   B. 6   C. 7   D. 8

(4) 以下不属于螺纹规塞的是( )螺丝塞规。
    A. 普通粗牙   B. 细牙   C. 圆孔塞规   D. 管子螺纹

(5) 下列选项中,( )指令不被用于螺纹加工。
A. G32　　　　B. G92　　　　C. G76　　　　D. G81

4. 简答题

(1) 简述丝锥分类。

(2) 简述加工循环指令 G74 指令和 G84 指令的功能和区别。

(3) 简述螺纹检测方法。

自我学习检测评分表如表 6-5 所示。

表 6-5　自我学习检测评分表

项　目	目 标 要 求	分值	评 分 细 则	得分	备注
学习关键知识点	(1) 了解常见丝锥的分类及应用 (2) 掌握攻丝加工常见的几种工艺方法 (3) 掌握螺纹攻丝的操作方法 (4) 掌握攻丝指令 G84 指令和 G74 指令的使用,理解各参数的含义 (5) 掌握螺纹孔的正确测量方法	25	理解与掌握		
工艺准备	(1) 能够正确识读零件图 (2) 能够独立确定加工工艺过程,并正确填写螺纹孔加工工艺文件 (3) 能根据工序加工工艺,编写正确的加工程序	25	理解与掌握		
上机训练	(1) 能够根据零件结构特点和精度合理选用量具,并正确、规范地测量相关尺寸 (2) 掌握螺纹加工刀具的选择与使用方法 (3) 掌握驱动轴、从动轴螺纹孔加工的操作流程 (4) 能够正确操作数控加工中心,并根据加工情况调整加工参数	50	(1) 理解与掌握 (2) 操作流程		

思政小课堂

Projects Guidance

CNC machining technology has been widely used in the manufacturing industry at home and abroad. The rapid development of the CNC machining industry also puts forward higher requirements for CNC programming and operation technicians. Guided by practical engineering projects and oriented by practical application, the scientific inquiry ability and problem-solving ability of students can be better cultivated. The Geneva mechanism is widely used in the machinery industry to realize intermittent motion. It has the characteristics of simple mechanism, reliable operation, and high transmission efficiency. In this book, the relevant parts of the typical balance wheel intermittent motion mechanism (Figure 0-1) are taken as the carrier, and six typical projects (Figure 0-2) are taken as the main line. Focus on learning the professional knowledge and operation skills of CNC turning and milling, such as process formulation, CNC programming and operation.

Figure 0-1   Typical balance wheel intermittent motion mechanism

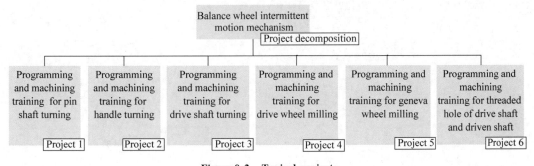

Figure 0-2   Typical projects

The balance wheel intermittent motion mechanism contains 6 typical parts and several standard components. The professional knowledge and operation skills required for CNC machining of these 6 parts are studied and trained respectively to complete the machining of the parts and finally complete the assembly of the mechanism. Each project involves the learning of the specialized knowledge and skills required to complete the machining of parts, such as equipment, fixtures, cutters, basic instructions, and machine operations. According to the technological design of CNC machining, targeted learning and training are carried out to master the technological design, CNC programming and machine operation of CNC turning and milling.

Project 1 Programming and Machining Training for Pin Shaft Turning

➢ Mind map

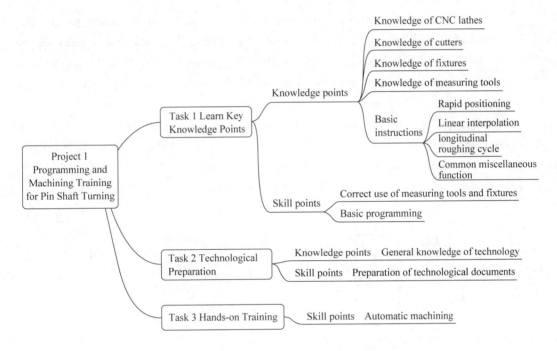

➢ Learning objectives

Knowledge objectives

(1) Be able to recognize drawings of shaft parts.

(2) Know the use and characteristics of multi-diameter shafts.

(3) Understand the meaning of each parameter of the longitudinal machining cycle instruction.

Ability objectives

(1) Master the selection and installation methods of external turning cutters and cut-off tools.

(2) Be able to independently determine the process routine and fill in the technological documents correctly.

(3) Be able to operate the CNC lathe correctly and adjust the machining parameters

according to the machining conditions.

(4) Be able to select measuring tools reasonably according to the accuracy and the structural characteristics of parts and be able to measure the relevant dimensions correctly and normatively.

Literacy goals

(1) Develop students'scientific spirit and attitude;

(2) Cultivate students'engineering awareness;

(3) Develop students'teamwork skills.

➤ Task introduction

The main function of the multi-diameter shaft is to locate the installed parts. The shaft shoulders with different heights can limit the movement or movement trend of the parts on the shaft along the axis direction, thereby preventing the installed parts from slipping during work, and reducing the influence of the axial pressure generated by some parts on other parts. Therefore, it is widely used in production and life.

According to the requirements of the part drawing Figure 1-1, the processing technology and the CNC machining program are developed, and then the machining of the pin part is completed. As a typical shaft part, it is made of 45 steel, and its blank is quenched and tempered, and the surface is required to be smooth without scratches.

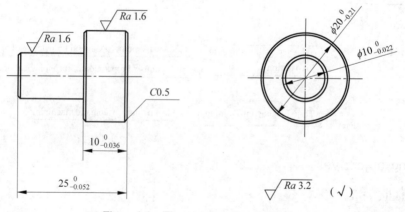

Figure 1-1　The part drawing of the pin

Task 1　Learn Key Knowledge Points

1.1　Preliminary understanding of CNC lathes

The CNC lathe is one of the most widely used CNC machine tools. It is mainly used for cutting inner and outer cylindrical surfaces of shaft parts or disc parts, inner and outer conical surfaces of arbitrary taper angles, complex rotating inner and outer surfaces and cylindrical and conical threads, etc. It can also perform the technologies, such as grooving, drilling, hole expanding, reaming, boring, etc.

The CNC machine tool automatically processes the parts to be processed according to the pre-programmed processing program. According to the instruction code and program format specified by the CNC machine tool, the programmers compiled the machining process route, process parameters, tool path, displacement, cutting parameters and auxiliary functions of the part into a machining program list, and then recorded the contents of this program list on the control medium, and then input them into the numerical control device of the CNC machine tool, so as to control the machine tool to realize the machining of the parts.

A CNC lathe is composed of the CNC equipment, the lathe bed, the headstock, the tool-head feed system, the tailstock, the hydraulic system, the cooling system, the lubrication system, the chip conveyor and other parts. CNC lathes can be divided into two types: vertical CNC lathes and horizontal CNC lathes, as shown in Figure 1-2. Vertical CNC lathes are used for turning disc parts with large turning diameters. Horizontal CNC lathes are used for turning disc parts with long axial dimension or small disc parts.

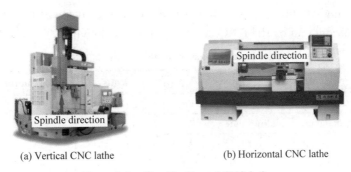

(a) Vertical CNC lathe　　　　(b) Horizontal CNC lathe

Figure 1-2　Classification of CNC lathes

The CNC lathe CK6150e (as shown in Figure 1-3) adopts the flat bed structure. Its spindle structure adopts the typical structure of front and rear two-point support, and the

Figure 1-3　The CNC lathe CK6150e

spindle has a high stiffness. The main drive is regulated by three-speed stepless variable frequency, and the speed range is 45~1600r/min. The main drive gear pairs are all hardened and grinded, and all transmission countershaft and rolling bearings are lubricated by powerful oil, so it has high-speed and low-temperature lift performance. The design of the headstock box fully considers the heat dissipation measures and shock absorption mechanism, so that the headstock has the characteristics of low noise and high transmission accuracy.

The spindle of the CK6150e drives the spindle box by the speed-change mechanism, which is generated by the variable frequency motor through the V belt. Through the frequency conversion system control of variable frequency motor, it can realize a manual three-speed stepless speed regulation, with a range of 100~2000r/min (forward and reverse rotation). The precision ball screw assembly is driven by the actuating motor to realize the transverse (X-axis) and longitudinal (Z-axis) rapid movement and feed movement. The slide lead rail is affixed with anti-crawling plastic soft belt, which can ensure the positioning accuracy and repetitive positioning accuracy of the CK6150e. The CK6150e is equipped with a manual three-jaws self-centering chuck and a four-station CNC tool holder, so it has the characteristics of high positioning accuracy, stability and reliability, wide application range, simple structure, and convenient maintenance.

The main parameters of the CNC lathe CK6150e is shown in Table 1-1.

Table 1-1 Main parameters of the CNC lathe CK6150e

Items	Units	Technical specifications
Maximum turning diameter of lathe bed	mm	$\phi 500$
Maximum turning diameter of sliding plate	mm	$\phi 300$
Maximum workpiece length between two center points	mm	880
Spindle through-hole diameter	mm	$\phi 80$
Spindle speed range	(r/min)	100~2000
Spindle motor power	kW	7.5(frequency conversion)
CNC tool holder		vertical tool holder with four operation station
X/Z-axis stroke	mm	310/880
X/Z-axis rapid return traverse speed	(mm/min)	8000/10000
X/Z-axis feed speed range	(mm/min)	1~2000
Tailstock sleeve diameter	mm	$\phi 75$
Tailstock sleeve stroke	mm	150
Countersink taper of tailstock sleeve		Morse 5
Dimensions (length× width× height)	mm	2900×1500×1800
Net weight	kg	2800
CNC system		FANUC 0i TF

1.2 Turning features

Turning is a cutting method that takes the rotation movement of the workpiece as the main movement and the tool holder as the feed movement. Its characteristics are:

(1) Turning is relatively efficient and has higher efficiency than milling and grinding. Turning can adopt higher revolution speed of workpiece and large cutting depth (back engagement of cutting edge) to achieve efficient machining.

(2) Turning has the characteristics of high production efficiency and wide processing range. Different cutters can be used to complete machining tasks such as internal and external cylindrical surface, end surface, conical surface, groove (rectangular groove, arc groove and irregular groove) and thread.

(3) Turning has high machining accuracy. Its economical machining precision is generally IT9~IT7, and the value of surface roughness Ra is generally 12.5~1.6μm. The precision turning accuracy can up to IT5 and the value of surface roughness Ra up to 0.20μm.

As shown in Figure 1-4, the cutting speed in turning is usually represented by the spindle speed n. The spindle speed (n) is the number of revolutions per minute of the workpiece on the spindle, and its unit is r/min(rpm). The feeding speed is usually expressed by the relative movement distance f_n between the workpiece and the cutter per revolution, and its unit is mm/r. The cutting depth can be divided into axial cutting depth and radial cutting depth, which are expressed by a_p, and the unit is mm.

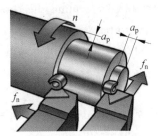

Figure 1-4 Turning model

1.3 Knowledge of cutters

Cutters is crucial for CNC turning. The selection of the types, materials and angles of the cutter directly affects the dimensional accuracy, surface quality and service life of the machined parts.

1.3.1 Classification of cutters

According to the different manufacturing materials, cutters can be divided into high-speed steel cutters, carbide-tipped cutting tools, diamond cutters, cubic boron nitride cutting tools, ceramic cutting tools and coated cutting tools, etc.

There are many types of turning tools, which can be generally divided into integral cutters, tipped tools, mechanically clamped tools and indexable turning tools according to the structure. According to the different processing purposes, it can be divided into cylindrical turning tool, boring cutter, slotting tool, external threading tool, etc., as shown in Figure 1-5.

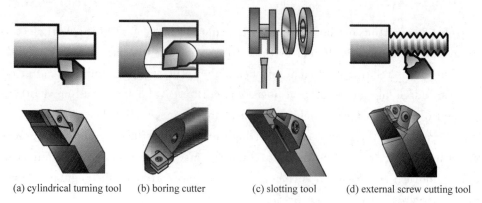

(a) cylindrical turning tool　　(b) boring cutter　　(c) slotting tool　　(d) external screw cutting tool

Figure 1-5　Commonly used turning tools

1.3.2　Cutters selection

The selection of cutters is crucial for CNC turning. The selection steps are as follows:

(1) Determine the processing type. According to the position characteristics of the parts to be processed and the processing strategy, the processing type such as roughing or finish machining, inner or outer contour processing (external turning, face cutting, profile turning and groove processing, etc.) is clarified.

(2) Determine the tool type. According to the content of working procedure, select the appropriate tool type, such as cylindrical turning tool, boring cutter, slotting tool, threading tool, etc.

(3) Determine the shank size. Select shank size according to machine tool requirements and aperture size.

(4) Choose a blade. According to the contour features and processing requirements, determine the shape, model, groove type and nose radius of the blade and the cutter brand. The mechanically-clamped indexable cutter is often used in CNC turning. The choice of a cutter is mainly the choice of a blade, that is, the choice of material, size, shape, and nose arc radius of the blade. The blade material is mainly determined according to the material of the workpiece to be processed, the accuracy requirements of the workpiece surface, the size of the cutting load, and whether there is shock and vibration during the cutting process. The size of the blade mainly refers to the length of the effective cutting edge, which is determined according to the cutting depth, a_p, and the cutting edge angle, k_r. You can refer to the relevant manual when using it. The shape of the blade is selected according to factors such as the surface shape of the workpiece to be processed, the cutting method, the cutter' service life, and the number of indexable of the blade. The size of the nose radius directly affects the strength of the tool nose and the surface roughness of the machined part. Usually, in the case of finishing machining with a small

cutting depth, slender shaft machining or poor machine tool stiffness, a smaller nose radius is selected. In the case of roughing with high strength of cutting edge and large diameter of part, a larger nose radius is selected.

1.4 Knowledge of fixtures

Fixture, which is also known as jig, refers to the device used to fix the processing object in the process of mechanical manufacturing, so that the workpiece occupies the correct position to accept construction or inspection. In a broad sense, in any working procedure in the process, the device used to install the workpiece quickly, easily and safely can be called a fixture. The fixture is usually composed of a locating component (determining the correct position of the workpiece in the fixture), a clamping device, a tool setting guide component (determining the relative position of the cutter and the workpiece or guiding the cutter direction), a indexing device (enabling the workpiece can complete the machining of several stations in a single installation, including rotary indexing device and linear moving indexing device), connecting elements and clamping body (clamp base) and so on.

The most common fixture for lathes is a chuck, which is mainly used to machine various forming surfaces such as end face, outer circle, inner hole and thread on the parts. It is one of the most common fixtures. According to the different power used to drive the claw, the chucks are divided into hand chucks and power chucks. A hand chuck consists of the chuck body, the movable claw, and the claw driving mechanism. Commonly used are three-jaw self-centering chuck and four-jaw chuck in which each jaw can move independently. Below the guide parts of the hand three-jaws self-centering chuck, there are threads that mesh with the plane threads on the back of the dishing bevel gear. When the wrench rotates the bevel gear through a square hole, the dishing gear rotates. At the same time, the flat thread on the backside drives the three jaws close to or away from the center to clamp workpieces of different diameters. Reverse jaws can also be used for mounting workpieces with larger diameters. Power chucks mainly include hydraulic chucks and pneumatic chucks. Figure 1-6 shows some common fixtures for CNC lathes.

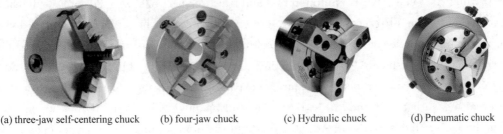

(a) three-jaw self-centering chuck   (b) four-jaw chuck   (c) Hydraulic chuck   (d) Pneumatic chuck

Figure 1-6 General fixtures for CNC lathes

1.5　Knowledge of measuring tools

　　Measuring tool is the abbreviation of physical measuring tool, which is an instrument with a fixed shape when used to reproduce or provide one or more known quantities of a given quantity. Measuring instruments commonly used in machining include standard instruments, common instruments and special instruments. Standard appliances refer to measuring tools used as measurement or verification standards, such as measuring blocks, polygonal prisms, surface roughness comparison samples, etc. Common instruments are also known as universal measuring tools, generally refer to general-purpose measuring instruments uniformly manufactured by measuring tool factories, such as rulers, plane-tables, angle rulers, calipers and micrometers, etc. Special instruments are also known as non-standard measuring tools, refer to special measuring tools designed and manufactured to detect a certain technical parameter of the workpiece, such as internal and external groove calipers, wire rope calipers, step gauges, etc. The measuring tool is a instrument that reproduces the measuring value in a fixed form. Its characteristics are as follows:

　　(1) It directly reproduces the unit value, that is, the nominal value of the measuring tool is the actual size of the unit value. For example, the gauge block itself reproduces the unit of length.

　　(2) Generally, there is no measuring mechanism, no indicator or moving component. For example, the gauge block is only a physical object that reproduces the unit value.

　　(3) Since there is no measurement mechanism, the measured value cannot be directly measured without relying on other equipped measurement instruments. For example, the gauge block should be equipped with interferometers and optical meters, so it is a passive measurement instrument.

1.5.1　Vernier caliper

　　The most common measuring tool in turning is the vernier caliper, as shown in Figure 1-7, which is a measuring tool for measuring length, inner and outer diameter, and depth. The vernier caliper consists of a main ruler and a vernier which can slide on the main ruler. The main ruler is generally in millimeters, and the vernier has 10, 20 or 50 divisions. According to the different divisions, vernier calipers can be divided into ten-division vernier calipers, twenty-division vernier calipers, and fifty-division vernier calipers. Their corresponding lengths are 9mm, 19mm and 49mm respectively. On the main ruler and the vernier of the vernier caliper, there are two pairs of active measuring jaws, namely the internal measuring jaw and the external measuring jaw. The internal measuring jaw is usually used to measure the inner diameter, and the internal measuring jaw is usually used to measure the length and the outer diameter.

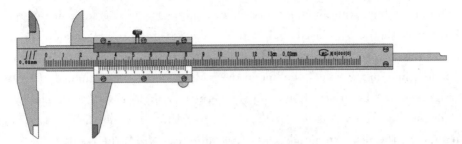

Figure 1-7 Vernier caliper

Vernier caliper is a commonly used instrument for measuring length in industry, which consists of a ruler and a vernier that can slide on the ruler. When viewed from the back, the vernier is a whole. There is a spring piece between the vernier and the ruler, and the elastic force of the spring piece is used to make the vernier and the ruler close together. There is a fastening screw on the upper part of the vernier, which can fix the vernier at any position on the ruler. Both the ruler and the vernier have measuring jaws. The internal measuring jaw can be used to measure the width of the groove and the inner diameter of the pipe, and the external measuring jaw can be used to measure the thickness of the part and the outer diameter of the pipe. The depth gauge can connect with the vernier gauge to measure the depth of grooves and cylinders. There are scales on the ruler and vernier. Taking a vernier caliper with an accuracy of 0.1mm as an example, the minimum division on the ruler is 1mm, and there are 10 small equal division scales on the vernier, with a total length of 9mm, that is, each division is 0.9mm, which is 0.1mm different from the minimum scale on the ruler. When the measuring jaws are close together, the zero scales of the ruler and the vernier are aligned. The difference between their first scale line is 0.1mm, the difference between the second scale line is 0.2mm, ..., the difference between the tenth scale line is 1mm, that is, the tenth scale line of the vernier is exactly aligned with the 9mm scale line of the ruler. When the length of the object measured between the jaws is 0.1mm, the vernier should move 0.1mm to the right. At this time, its first scale line is just aligned with the 1mm scale line of the ruler. Similarly, when the fifth scale line of the vernier is aligned with the 5mm scale line of the ruler, it means that there is a width of 0.5mm between the two measuring jaws, and so on. When measuring a length greater than 1 mm, the integral part of millimeters should be read from the scale line on the ruler opposite to the zero scale of the vernier.

First, read integer part in millimeters on the ruler based on the zero scale of the vernier. Then observe which scale line on the vernier are aligned with the scale line of the ruler. For an example, if the sixth scale line of the vernier is aligned with the scale line of the ruler, the fractional part is 0.6mm (if there is no exactly aligned scale line, take the closest alignment scale line for reading). If there is zero error, the above result will be subtracted from zero error (zero error is negative, which is equivalent to adding zero error of the same value), and the reading result is: L = integer part + fractional part-zero error. To judge which scale line on the vernier is aligned with the scale line of the ruler, the following method can be used: select three adjacent lines, if the line on the left is to the right of the corresponding line on the ruler, and the line on the right is to the left of corresponding line on the ruler, the line in the middle is considered aligned. $L=$ the scale before alignment $+$ the nth scale line on the vernier is aligned with the scale line on the ruler $\times$ division value. If you need to measure several times to get the average, you don't need to subtract the zero error each time, just subtract it from the result.

The following is an example of a state of a vernier caliper with an accuracy of 0.02mm shown in Figure 1-8.

(1) Read the scale value on the ruler that to the left of the zero scale of the vernier. It is the integer part of the final reading. As shown in Figure 1-8, the integer part is 33 mm.

(2) There must be a scale line on the vernier aligned with the scale line of the ruler. Read the scale line on the vernier from the zero scale. As shown in Figure 1-8, the number of grids on the left is 12, multiply the accuracy of the vernier caliper by 0.02mm, then get the fractional part of the last reading. Or directly read the value on the vernier. As shown in Figure 1-8, it is 0.24mm.

(3) Add the integer and fractional parts to get a final value of 33.24mm.

Figure 1-8 Vernier caliper reading

1.5.2 Micrometers

A micrometer, sometimes known as a micrometer screw gauge, is widely used as a universal measuring tool for CNC turning part measurement. It is divided into mechanical micrometer and digital micrometer according to the display mode, as shown in Figure 1-9. According to the application, it is divided into outside micrometer, inside micrometer, depth micrometer, thread micrometer, tube micrometer and so on. Among them, the outside micrometer, especially the mechanical outside micrometer is the most common. As shown in Figure 1-10, it is mainly composed of the frame, the anvil, the spindle, the sleeve, the thimble and the ratchet stop. It has the characteristics of low cost, accurate measurement, stable and reliable use, etc. It is mainly used to measure the outer diameter and length of parts.

When using a micrometer to measure the length of a part, one hand should control the stability of the frame, and the other hand should rotate the ratchet stop to make the measuring end of the micrometer fully fit the measured part. The reading for micrometer is divided into three steps. First, take the left end face of the thimble as the reference line,

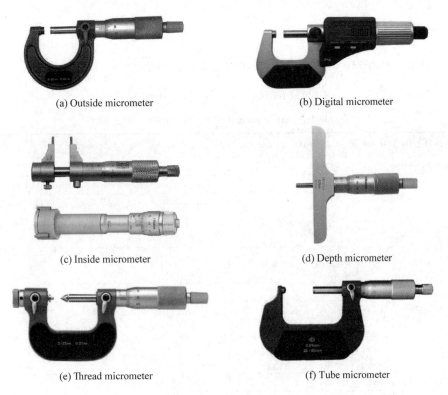

Figure 1-9 Micrometers

read the graduation value of the scale line on the sleeve (read-only integer). Then use the longitudinal line on the sleeve as the reference line, read the scale value on the thimble (0.01mm division). Finally, estimate the thousandth value (0.001mm) according to the actual position of the thimble. If the scale line on the thimble is aligned with the reference line, the thousandth value is 0. The three values are added to obtain the final value.

As shown in Figure 1-11, the reading steps are as follows: first, read the value of the sleeve as 5.5mm, then read the value of the thimble as 0.04mm, and finally estimate the thousandth value to be 0.003mm. Thus, the reading is given by adding the three values, which is 5.543mm.

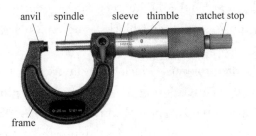

Figure 1-10 Micrometer structure

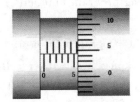

Figure 1-11 Micrometer reading

1.6　Basic instructions

1.6.1　Basic format of the CNC turning program

The basic format of a CNC turning program is shown in Table 1-2.

Table 1-2　Basic format of a CNC turning program

Program details	Number	Instruction	Remark
Program header	①	O0001;	Program name
Program body	②	T0101;	Call the cutter and cutter compensation
	③	G97 G99 S800 M03;	The spindle speed is in r/min. The feed speed is in mm/r. The spindle is forward at a speed of 800r/min
	④	G0 X__ Z__ M8;	Quickly move to point (X__ Z__) and open the coolant
	⑤	... ...	Blocks for turning
	⑥	G0 X__ Z__;	Back to the safe location
	⑦	M5;	Spindle is stopped
	⑧	M9;	Shut-off the coolant
	⑨	...	If other tools are needed, repeat the steps ②~⑧
Program end	⑩	M30;	End the program

The program head generally contains the program starter, the program name or program number. The program starter can be %, :, MP and other different identifiers according to different CNC systems. It can also be omitted in some CNC systems. Therefore, it needs to be specified according to programming manual of the machine tool when programming. In the FANUC 0i system, the program header can be directly represented by a program number with a capital letter "O" followed by a number. The program number is a positive integer with a value range of 0001~9999. Usually, programs in the range of O9000~O9999 are locked, which are used to save the specific function program of the machine tool. The zero before the number can be omitted when inputting program number. The program consists of several blocks, and each block ends with ";".

The program body is the main part of the program, which is the process of turning the workpiece with cutters. If there is only one cutter during the machining, the program body is the steps ②~⑧. If there are multiple cutters, repeat the process ②~⑧. Among them, ②~④ are the initialization of the turning process. In these steps, the cutter and cutter compensation are called, and the units of spindle revolution speed and feed speed are set, and the spindle is started (pay attention to the rotation direction of the spindle). Then control the cutter to move quickly to the safe position in preparation for formal turning. Step ⑤ is the formal turning process, which represents the turning process of an area on the workpiece. If an area needs to be turned multiple times, it is only necessary to repeat the process of "feeding, turning and back-off". If turning multiple areas, just repeat ⑤~⑥. During CNC lathe machining, after all areas are cut, the tool should be retracted to the safe position to facilitate the part measurement and loading and unloading, and to prevent collisions. The spindle and coolant also should be shut off.

The program end contains the end directive and the end identifier of the program. The commonly used end directive in the FANUC system is M02 or M30 (usually M30). Similar to program header, the end identifier should be specified according to programming manual of the machine tool.

Commonly used preparatory function instructions in FANUC system are shown in Table 1-3.

Table 1-3 Commonly used preparatory function instructions

Instructions	Usage	Instructions	Usage
G00	Rapid positioning	G42	Right compensation of the corner radius
G01	Line interpolation	G70	cycle instruction of fine machining
G02	Clockwise circular interpolation	G71	Longitudinal roughing cycle
G03	Counterclockwise circular interpolation	G72	Lateral roughing cycle
G04	Program stop	G73	Profiling roughing cycle
G20	Imperial input	G92	Thread machining cycle
G21	Metric input	G96	Constant linear speed control
G28	Return to reference point	G97	Constant revolution speed control
G40	Cancel corner radius compensation	G98	Feed-rate per minute
G41	Left compensation of the corner radius	G99	Feed-rate per revolution

1.6.2 Basic preparatory function instructions

1. Rapid positioning instruction G00

The instruction G00 enables the cutter to rapidly move from the current point to the next positioning point commanded by the block at the maximum feed speed preset by the CNC system (The actual running speed needs to consider the position of the rapid override knob). As shown in Figure 1-12, there may be various motion trajectories during the rapid positioning process. The common trajectory is to locate from point A to point C, and then move from point C to point B. Due to the different parameter settings of the machine tool when it leaves the factory, the trajectory is also different. When operating the machine tool, you should pay attention to its

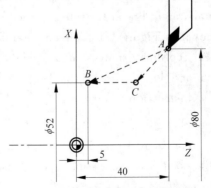

Figure 1-12 Rapid positioning

movement trajectory to avoid collision accidents. The format of the quick positioning command is the command code followed by the position coordinates of the positioning end point.

There are two ways to specify the movement of the cutter, that is, the absolute coordinate and the relative (incremental) coordinate. The absolute coordinate command is to specify the end point coordinate for the moving position of the cutter by the method of the actual coordinate value of the workpiece coordinate system. During programming, the radial and axial directions are represented by addresses X __ and Z __ respectively. The relative coordinate command is also called the incremental coordinate command. It is a method of specifying the end point coordinate in the form of increment relative to the current point of the cutter. The radial and axial directions are represented by addresses U __ and W __ respectively. Absolute coordinates and incremental coordinates can be mixed in the same block.

Note: During CNC turning programming, in order to maintain the dimensional consistency of marking, programming and measurement, the radial coordinates are usually represented by diameter values, that is, diameter programming. Although it can also be represented by a radius value by modifying the machine parameters, it is rarely used.

Instruction format:

G00 X __ Z __ ;

For example, the instruction to quickly move from point A to point B is:

G00 X52 Z5;
or G00 U－28 W－35;
or G00 x52 W－25;
or G00 U－28 Z5;

It is recommended that beginners try to use absolute coordinate instruction to reduce unnecessary mistakes.

2. Linear interpolation instruction G01

The linear interpolation instruction, G01, is to move the cutter from its current position to the position required by the instruction in a straight line at the rate specified by the F code, shown in Figure 1-13. In an interactive manner, each coordinate moves in a straight line with an arbitary slope accord to the feed rate F.

Instruction format:

G01 X __ Z __ F __ ;

For example, the instruction to move from point A to point B at the specified speed F is:

G01 X52 Z5 F0.2;

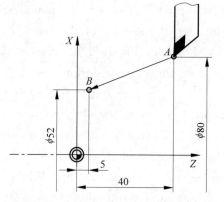

Figure 1-13　Linear interpolation

Or

```
G01 U-28 W-35 F0.2;
```

3. Longitudinal rough turning cycle instruction G71

The instruction G71 only needs to specify the amount of cutting depth, the amount of retraction, the machining allowance and the finishing route, and the system can automatically give the roughing route and the number of machining times to complete the rough machining of the inner and outer circular surfaces. Shown in Figure 1-14, instruction format is as follow:

```
G71 U (Δd) R (e);
G71 P (ns) Q (nf) U (Δu) W (Δw) F(f);
```

Δd: It indicates the amount of cutting depth each time, which is generally specified by the radius value. For example, 45 steel parts generally take 1~2mm, and aluminum parts generally take 1.5~3mm.

e: It represents the amount of retraction in X direction, specified by the radius value, which generally is 0.5~1mm.

ns: It indicates the number of the starting block in the roughing contour block.

nf: It indicates the number of the ending block in the roughing contour block.

Δu: It indicates the roughing allowance in the X direction, which is negative when machining the inner contour.

Δw: It indicates the roughing allowance in the Z direction, generally take 0.05~0.1mm.

Instruction action: First, the cutter rapid feed from the cycle starting point, then turns, and then retracts by 45°, and finally returns quickly. First feed, then cut, then retract at 45°, and finally return quickly, and so on to complete roughing, as shown in Figure 1-14.

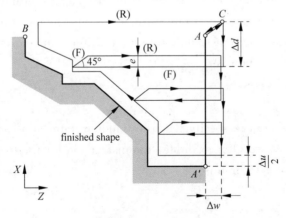

Figure 1-14 Longitudinal roughing cycle instruction G71

Precautions:

(1) When using G71 instruction for rough machining, the F and S included in the block ns~nf are invalid for the rough turning cycle.

(2) Subprograms cannot be called in the block whose sequence number is ns~nf.

(3) Normally, this instruction is mainly used for contour machining with monotonic increase or decrease in X-axis and Z-axis directions.

(4) Usually, the first sentence in the block ns~nf must move along the X direction (FANUC 0i-mate system).

4. Finishing process cycle instruction G70

Instruction G70 is used to remove finishing allowance.

Instruction format:

G70 P(ns) Q(nf) F(f);

It is used for finishing, cutting off the machining allowance left after rough machining by the instruction G71.

Precautions:

(1) The F and S instructions in the block ns~nf are valid.

(2) The cutter returns to the starting point of the cycle after turning by the instruction G70.

5. Spindle speed control instruction

Spindle speed control instruction includes Constant linear speed control instruction G96 and Constant rotative speed control instruction G97.

G96 instruction is used to specify that the spindle rotates at a constant linear speed, and is mainly used for machining contours with large changes in radial dimensions, so that the surface quality of the machined parts tends to be consistent. That is, the spindle speed changes with the change of the cutting diameter. It can be seen from the formula $n = (1000D)/(\pi d)$ that the closer the tool is to the center, the higher the spindle speed will be. In order to prevent the spindle speed from increasing continuously, it is usually necessary to use the instruction G50 to limit the maximum spindle speed. For example, the program to set the linear speed of spindle rotation to 120m/min and the maximum rotative speed not to exceed 2000r/min is:

G50 S2000;
G96 S120;

G97 instruction is used to specify that the spindle rotates at a constant rotative speed and is the most widely used. For example, the instruction for the forward rotation of the spindle with a speed of 960r/min is:

G97 S960 M3;

6. Cutting speed control instruction

Cutting speed control instruction includes the following.

(1) Feed speed per minute: G98 instruction is used to set the cutter feed speed. Its unit is mm/min.

(2) Feed mount per revolution: G99 instruction is used to set the cutter feed mount.

Its unit is mm/r.

The relationship between the two is

$$F_m = F_r \cdot n$$

Where F_m is the feed speed per minute, F_r is the feed amount per revolution, and n is the spindle rotative speed.

Note: The actual movement speed of the machine tool is also related to the feed ratio.

1.6.3 Miscellaneous function M-codes

Miscellaneous function M-codes are used to control some miscellaneous actions of the machine tool, such as start or stop the spindle and coolant on or off, etc. The commonly used Miscellaneous function M-codes in FANUC system are shown in Table 1-4.

Table 1-4 Commonly used miscellaneous function M-codes in FANUC system

Number	Code	Function	Remark
1	M00	Compulsory stop	
2	M01	Optional stop	Machine only stops at M01 if operator pushes the optional stop button.
3	M02	End of Program	
4	M03	Start Spindle-Clockwise	
5	M04	Start Spindle-Counter Clockwise	
6	M05	Stop Spindle	
7	M08	Flood coolant on	
8	M09	Coolant off	
9	M30	Program reset & rewind	

Note: Usually only one M-code is allowed in each block.

1.6.4 CNC lathes utilize T-code

The T-code is used to call the cutter and the cutter compensation. The format is:

T __

The cutter function instruction consists of the letter T and 4 digits. The first two digits represent the cutter number, and the last two digits represent the cutter compensation number. Usually, the cutter number and the cutter compensation number are the same.

1.6.5 spindle function S-code

CNC lathes use S-code to specify the spindle slewing speed of the machine tool, whose unit is r/min or m/min. The value specified by S code is a positive integer. When the value range is greater than the maximum rotative speed of the machine tool, the spindle rotates at the maximum rotative speed. The code format is:

S __

Miscellaneous function M-codes control the start and stop of the spindle. M03 is the forward rotation of the spindle, M04 is the reverse rotation of the spindle, and M05 is the spindle stop. Observed against the positive direction of the Z-axis, the counterclockwise rotation of the spindle is defined as forward rotation.

For example, the instruction for the forward rotation of the spindle with a speed of 800r/min is:

S800 M3;

Task 2   Technological Preparation

1.7 Parts drawing analysis

According to the operation requirements of the part, 45 steel is selected as the blank material of this part, and the blanking dimension is set as $\phi 25 \times 60$. Using the excircle of the blank with a diameter of $\phi 25$ as the rough reference, the end face and excircle of the multi-diameter shaft with diameters of $\phi 10$ and $\phi 20$ are roughed, and then finished to ensure the dimensional accuracy and surface quality, and finally cut the parts to ensure the length requirements.

The two excircles of the multi-diameter shaft need to be processed at the same time in the case of one clamping to ensure better concentricity. If the positioning is divided into two times, the positive difficulty of the part is relatively high, and it is not easy to ensure the working needs of the part.

Note that when clamping the blank, attention should be paid to the extended length of the bar to avoid collision between the cutter and the chuck.

1.8 Technological design

According to the analysis of the part drawing, the technological process is designed as shown in Table 1-5.

Table 1-5 Technological process card

Machining process card	Product model	CLJG-01		Part number	BX-01	Page 1	
	Product name	Geneva mechanism		Part name	Pin	Total 1 page	
Material grade	C45	Blank size	$\phi 25 \times 60$	Blank quality	kg	Quantity	1
Working procedure				Work section	Technical equipment	Man-hours/min	
No.	Name	Content				Preparation & conclusion	Single piece
5	Preparation	Prepare the material according to the size of $\phi 25 \times 60$		Outsourcing	Sawing machine		

10	Turning	Using the excircle of $\phi 25$ as the rough reference, turn the excircle and the end face of $\phi 10$ and $\phi 20$	Turning	Lathe, outside micrometer	45	30
15	Cleaning	Clean the workpiece, debur sharp corner	Locksmith			5
20	Inspection	Check the workpiece dimensions	Examination			5

Based on the 10th working procedure of turning, this training task is designed, and the corresponding working procedure card is formulated as shown in Table 1-6.

Table 1-6 Working procedure card for turning

Machining working procedure card	Product model	CLJG-01	Part number	BX-01	Page 1
	Product name	Geneva mechanism	Part name	Pin	Total 1 page

Procedure No.	20	
Procedure name	Turning	
Material	C45	
Equipment	CNC lathe	
Equipment model	CK6150e	
Fixture	three-jaw self centering chuck	
Measuring tool	Vernier caliper	
	micrometer	
Preparation & Conclusion time	45min	
Single-piece time	30min	

Steps	Content	Cutters	$S/$ (r/min)	$F/$ (mm/r)	$a_p/$ mm	Step hours/min	
						mechanical	auxiliary
1	Workpiece installation						5
2	The outer surface and end face of $\phi 10$ and $\phi 20$ are roughed, and the finishing allowance is 0.2mm	Outer circle roughing turning tool	1200	0.2	1.5	15	
3	Finishing the outer surface and end face of $\phi 10$ and $\phi 20$	Outer circle finishing turning tool	1500	0.1	0.2	10	
4	Cutting and chamfering	Cut-off tool (width: 3mm) 700	700	0.07	0.1	10	
5	Dismantling and cleaning workpieces						5

1.9 CNC machining programming

According to the technology of the working procedure, the machining program is written as shown in Table 1-7.

Table 1-7 CNC program for pin turning

No.	Program Statement	Annotation
	O0001;	
N1	T0101;	Call the outer circle turning tool
	G97 G99 S1200 M03;	Set constant rotative speed control, unit of feed mount is mm/r, spindle speed is 1200r/min, forward rotation
	G0 X27 Z2 M8;	Rapidly locate to the beginning of the cycle ($X27$, $Z2$) and open the coolant
	G71 U1.5 R0.5;	Call the longitudinal roughing cycle instruction, the cutting depth is 1.5mm, and the retraction amount is 0.5mm
	G71 P10 Q20 U0.4 W0.05 F0.2;	Radial finishing allowance is 0.4mm (unilateral 0.2mm), the axial finishing allowance is 0.05mm, the feed rate is 0.2mm/r
N10	G0 X0;	
	G1 Z0;	
	X9;	
	X10 Z-0.5;	
	Z-15;	
	X19;	
	X20 Z-15.5;	
	Z-30;	
N20	X27;	
	G0 X100 Z150;	Quickly retract to ($X100$, $Z150$)
	M5;	Stop the spindle
	M9;	Coolant off
	M01;	Optional stop (the optional stop button needs to be pushed) to observe the completion of roughing
N2	T00101;	Call the cylindrical turning tool
	G97 G99 S1500 M3;	Set constant rotative speed control, unit of feed mount is mm/r, spindle speed is 1500r/min, forward rotation
	G0 X27 Z2 M8;	Rapidly locate to the beginning of the cycle ($X27$, $Z2$) and open the coolant
	G70 P10 Q20 F0.1;	Call the finishing cycle instruction with a feed of 0.1mm/r
	G0 X100 Z150;	Quickly retract to ($X100$, $Z150$)
	M5;	Stop the spindle
	M9;	Coolant off
	M01;	Optional stop (the optional stop button needs to be pushed) to observe the completion of roughing
N3	T0202;	Call the external grooving cutter (cutter width is 3mm)

Continued

No.	Program Statement	Annotation
	G97 G99 S700 M3;	Set constant rotative speed control, unit of feed mount is mm/r, spindle speed is 700r/min, forward rotation
	G0 X22 Z2;	Quickly locate to X22, Z2
	Z-29 M8;	Quickly locate to Z-29 and turn on the coolant
	G1 X19 F0.07;	Pre-grooving
	X21 F0.3;	Retract
	Z-27;	
	X19 Z-28 F0.07;	Machining C0.5 chamfer with the right nose of the cutter
	X0;	Cut-off. According to the site conditions, you can also choose to process to X2, and then manually break it
	G0 X100;	Radial retract
	Z150;	Axial retract
	M5;	Stop the spindle
	M9;	Coolant off
	M30;	End of the program

Task 3   Hands-on Training

1.10 Equipment and appliances

Equipment: CK6150e CNC lathe.
Cutters: the outer circle turning tool, the cut-off tool with a blade width of 3mm.
Fixture: the three-jaw self-centering chuck.
Tools: chuck wrenches, tool holder wrenches, etc.
Gauges: the 0~150mm vernier caliper and the 0~25mm outside micrometer.
Blank: $\phi 25 \times 60$.
Auxiliary appliances: cutter shim, brushes and so on.

1.11 Get to know the machine tool

1.11.1 Power on

Check whether there is any abnormality in various parts of the appearance of the machine tool (such as the scatter shield, the footplate, etc.). Check whether the lubricating oil and coolant of the machine tool are sufficient. Check whether there are foreign objects on the tool holder, the fixture, and the baffle plate of the lead rail. Check the state of each knob on the machine tool panel is normal. Check whether there is an alarm after power on the machine tool. Refer to Table 1-8 to check the machine state.

Table 1-8 Preparing card for machine start-up

Check Item		Test Result	Abnormal Description
Mechanical part	Spindle		
	Feed part		
	Tool holder		
	three-jaw self centering chuck		
Electrical part	Main power supply		
	Cooling fan		
CNC system	Electrical components		
	Controlling part		
	Driving section		
Auxiliary part	Cooling system		
	Compressed air system		
	Lubricating system		

If the inspection of the machine tool is normal, the power of the machine tool can be turned on by rotating the electrical switch at the back of the machine tool.

Rotate the emergency stop switch in the direction marked by the button to solve the emergency.

Press the "power on" button on the control panel to power on the CNC system. The system panel after power on is shown in Figure 1-15.

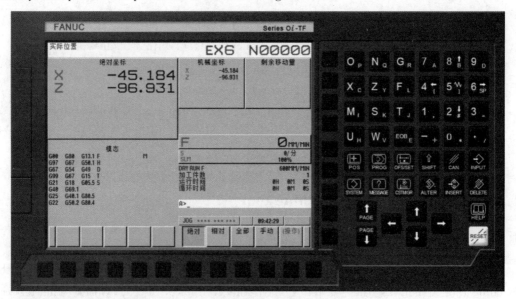

Figure 1-15 The CNC system panel after powering on

1.11.2 Operation panel

Before formally operating the machine tool, you should be familiar with the functions and operation methods of each button on the operation panel of the CNC lathe, and memorize the specific location of the emergency button.

1. Working mode selection keys

The working mode selection keys are shown in Figure 1-16. After a certain working mode is selected, there will be a corresponding mark on the display screen.

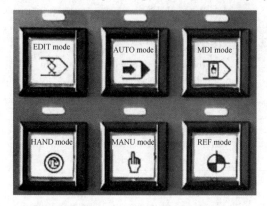

Figure 1-16 Working mode selection keys

Edit mode (EDIT) selection key: For editing programs or external CNC read-in.

Auto mode (AUTO) selection key: Used to automatically run programs that read into memory.

MDI mode: Used to run the program entered by the MDI panel of the controller.

Hand wheel mode (HAND) selection key: For moving the machine axes using an electronic hand wheel.

Manual mode (MANU) selection key: Use the direction keys on the operation panel to quickly move the machine axes.

Reference return mode (REF) selection key: Used for returning to the reference point.

2. Other function keys

Other common function keys on the operation panel are shown in Table 1-9. Due to different control systems, manufacturers, and factory batches, there will be some differences in the position and quantity of the panel function keys. Before operating the machine tool, you should be familiar with the operation manual of the current machine tool, and operate it in accordance with the requirements of safe operation.

Table 1-9 Common function buttons on the panel

Icon	Description
	Name: Program Start Button Select the machining program to be executed in the auto working mode (manual input, memory, online), and press the "Program Start" button to start executing the program

Continued

Icon	Description
	Name: Program Pause Button (1) In the auto working mode (manual input, memory, online), press the "Program Pause" button, and each axis immediately slows down and stops, entering a pause state (2) When the "Program Start" button is pressed again, the machining program resumes from the single block that is currently paused
	Name: Program Protection Switch (1) In order to prevent the program in the machine controller from being edited, cancelled, modified, or established by others, the key should be kept by a special person (2) In general, we set this key to the "OFF" position to ensure that the program is not modified or deleted (3) If we want to edit, cancel, or modify the program, this key should be set to the "ON" position
	Name: Single Block Button This function is only valid in the auto mode (1) When the light of this button is on, the function of executing a single block is valid. When this function is enabled, the program will be executed in a single block. After the current block is executed, the program will be suspended. After continuing to press the "Program Start" button, the program of the next block can be executed, and so on (2) When the light of this button is on, the function of executing a single block is invalid. The program will be executed until its end
	Name: Empty Run Button This function is only valid in auto mode (1) When the indicator light of this button is on, the Z-axis locking function is valid. When this function is turned on, the command to set the F value (cutting feed rate) in the program is invalid, and the movement rate of each axis is specified by the slow displacement rate (2) When the function is valid, if the program executes the cycle program, the feed rate cannot be changed by the slow feed rate or the cutting feed rate, and it keeps the F value in the control and displays at a fix feed rate
	Name: Skip Button This function is only valid in the auto mode (1) When the light of this button is on, the program selection skip function is valid. When this function is valid in the auto working mode, if a "/" (slash) symbol is specified at the beginning of a block, the block will be skipped and not executed (2) When the light of this button is off, the program selection skip function is invalid. When this function is invalid, even if there is a "/" (slash) symbol before a block, this block can be executed normally

Continued

Icon	Description
Optional Stop	Name: Optional Stop Button This function is only valid in the auto mode (1) When the light of this button is on, the program optional stop function is valid. When this function is valid, if there is an M01 command in the execution program, the program will stop at this block. To continue executing the program, press the program start button (2) When the light of this button is off, the program optional stop function is invalid. When this function is invalid, the program will not stop executing even if there is an M01 instruction in it
Machine Lock	Name: Machine Lock Button (1) When the light of this button is on, the machine lock function of all axes is valid. When this function is valid, no matter if any axis is moved in manual mode or automatic mode, the CNC will stop outputting pulses (movement commands) to the servo motor of this axis, but the command assignment is still performed, and the absolute and relative coordinates of the corresponding axis are also updated (2) The M, S, T code will continue to be executed and is not restricted by machine lock (3) After releasing this function, it is necessary to return to the mechanical zero point again. After the return to the reference point is correct and completed, other related operations can be performed. If the relevant operations are performed without returning to zero, it will cause coordinate offsets, and even abnormal phenomena such as collisions and program execution confusion, resulting in danger
F1	Name: F1 Button Reserved button. This button can be set according to the actual configuration of the machine tool. The operator cannot operate it
F2	Name: F2 Button Reserved button. This button can be set according to the actual configuration of the machine tool. The operator cannot operate it
F3	Name: F3 (Work Light Extension) Button Control the work light on or off without any operating mode restrictions
Spindle speed-down	Name: Spindle speed-down button (1) This button is located on the operation panel of the machine tool and is used to reduce the programmed spindle speed S. The actual speed = the value given by the programmed S command × the spindle speed reduction override (2) Used in conjunction with spindle function buttons

Continued

Icon	Description
Spindle speed-up	Name: Spindle speed-up button (1) This button is located on the operation panel of the machine tool and is used to increase the programmed spindle speed S. The actual speed = the value given by the programmed S command × the spindle speed reducing override (2) When the programmed speed exceeds the maximum speed of the spindle, and when the speed override reaches over 100%, the spindle speed is adjusted to the maximum speed of the spindle (3) Used in conjunction with spindle function buttons
Spindle Rotation-CW	Name: Spindle Rotation (Clockwise, CW) (1) After executing the S code once on this machine, select the manual mode, and press the "Spindle Rotation (CW)" button, the spindle will rotate clockwise. Spindle rotation speed = the previously executed spindle speed S value × the gear where the spindle tune knob is located (2) Conditions of use: ① Only available in Manual mode, Rapid mode, and Jog mode ② In auto mode, after executing the spindle rotation clockwise M03 command in the program, the indicator light of this button will be on ③ When Spindle Stop or Spindle Rotation (CCW) is valid, the indicator light turns off ④ The spindle must be stopped before you can specify a Spindle Rotation (CCW)
Spindle Stop	Name: Spindle Stop Button (1) Whether the spindle rotates clockwise or counterclockwise, pressing this button can stop the rotating spindle (2) Conditions of use: ① Only available in Manual mode, Rapid mode, and Jog mode ② Invalid in Auto mode (3) The indicator light will be on when the spindle is stopped, but if the "Spindle Rotation (CW)" or "Spindle Rotation (CCW)" is valid, the indicator light will be off
Spindle Rotation-CCW	Name: Spindle Rotation (Counterclockwise, CCW) Button (1) After executing the S code once on this machine, select the manual mode, and press the "Spindle Rotation (CCW)" button, the spindle will rotate counterclockwise. Spindle rotation speed = the previously executed spindle speed S value × the gear where the spindle tune knob is located (2) Conditions of Use: ① Only available in Manual mode, Rapid mode, and Jog mode ② In Auto mode, after executing the spindle rotation counterclockwise M04 command in the program, the indicator light of this button will be on (3) The indicator light will be on when the spindle rotates counterclockwise, but if the "Spindle Rotation (CW)" or "Spindle Stop" is valid, the indicator light will be off (4) The spindle must be stopped before you can specify a Spindle Rotation (CW)

Continued

Icon	Description
Cooling	Name: Cooling Button (1) In Manual, Rapid and Jog mode, when the indicator light is on by pressing this button, coolant is turned on (2) Press the "RESET" key and the cooling fluid stops spewing out, and the indicator light turns off (3) Pay attention to the direction of the coolant nozzle when the cooling fluid is turned on
Manual tool selection	Name: Manual tool selection Button In Manual, JOG and INC mode, each time this button is pressed, the cutter rotates one tool position in the plus direction
(↓)	Name: +X Control Button +X button: In the JOG mode, press and hold this button, the X-axis will move to the "+" direction (positive direction) of the machine X-axis according to the speed of the feed override/rapid override, and the indicator lamp will turn on at the same time. When the button is released, the axis stops moving in the "+" direction of X-axis, and the indicator lamp turns off at the same time In addition, this key is also used as the X-axis zero return trigger key
(↑)	Name: −X Control Button −X button: In the JOG mode, press and hold this button, the X-axis will move to the "−" direction (negative direction) of the machine X-axis according to the speed of the feed override/rapid override, and the indicator lamp will turn on at the same time. When the button is released, the axis stops moving in the "−" direction of X-axis, and the indicator lamp turns off at the same time
(→)	Name: +Z Control Button +Z button: In the Jog mode, press and hold this button, the Z-axis will move to the "+" direction (positive direction) of the machine Z-axis according to the speed of the feed override/rapid override, and the indicator lamp will turn on at the same time. When the button is released, the axis stops moving in the "+" direction of Z-axis, and the indicator lamp turns off at the same time In addition, this key is also used as the Z-axis zero return trigger key
(←)	Name: −Z Control Button −Z button: In the JOG mode, press and hold this button, the Z-axis will move to the "−" direction (negative direction) of the machine Z-axis according to the speed of the feed override/rapid override, and the indicator lamp will turn on at the same time. When the button is released, the axis stops moving in the "−" direction of Z-axis, and the indicator lamp turns off at the same time In addition, when the Z-axis negative direction movement command is executed in the program, the indicator light will also be on, and when the movement command is stopped, the indicator light will be off

Continued

Icon	Description
	Name: Overstroke Release Button (1) When the stroke of each axis exceeds the hardware limit, the machine tool will give an overstroke alarm and stop working at the same time. At this moment, press this button and use the hand-held unit to move the overstroke axis in the opposite direction in the handwheel mode (2) The absolute encoder machine tool does not need to press this button for overstroke
	Name: Manual Rapid Feeding Button This function is only valid in manual mode. Press this button in manual mode and its indicator light turns on. In this mode, the actual rapid feeding speed = the maximum speed of parameter setting G00 instruction × the override value ‰ where the rapid override key is located
	Name: Feeding Override and Feeding Override Adjusting (1) This knob is located on the operation panel of this machine tool to control the programmed G01 speed. Actual feeding speed = programmed given F command value × override value ‰ where the feed override switch is located (2) In the Jog mode, the function of this knob is to control the Jog feed override. The actual Jog feed speed = fixed value set by parameter × the override value ‰ where the feed override switch is located (3) Used in conjunction with the axes feed control keys
	Name: Rapid Override (1) These keys are located on the operation panel to control the programmed G00 instruction speed. Actual feeding speed = the maximum speed of parameter setting G00 instruction × override value ‰ where the rapid override key is located (2) In rapid movement, its function is to control the manual rapid feeding override. The actual rapid feeding speed = the maximum speed of parameter setting G00 × the override value ‰ where the rapid override key is located (3) Used in conjunction with the axes feed control keys

1.11.3 Return to reference point

If the incremental encoder is used for each feed axis of the CNC lathe, after the system is started, each axis needs to be manually returned to the reference point. First press the operation mode selection button , then press the +X button on the operation panel, and then press the +Z button to make each axis return to the reference point to determine the machine coordinate origin. When performing the operation of returning to the reference point, you should first return to the X-axis, and then return to the Z-axis to avoid collision. If the machine tool adopts an absolute encoder, there is no need to perform the operation of returning to the reference point after the machine is powered on.

1.12 Cutters preparation

Before machining, the cutters required for this task should be prepared. Table 1-10 is a list of the cutters required. Install the cutters correctly according to the sequence in the table.

Table 1-10 Cutters installation

Cutter No.	Toolbar	Blade	Installation tool
T1			
T2			

When installing the tool, the correct installation tool and method should be used. Wrong operation may damage the cutter and the toolbar, and even cause personal injury. The installation accuracy of the cutter also has a great impact on the machining accuracy and cutter life.

1.13 Set the workpiece origin

The process of workpiece origin setting is also called tool setting. Correct origin setting is very important for CNC machining. Tool setting can be divided into automatic tool setting and manual tool setting according to the degree of automation. Automatic tool setting requires equipment with automatic tool setting device, which is less used. Manual tool setting usually uses the trial cutting method, which is widely used.

1.13.1 Z-axis origin setting

Install the blank according to the processing requirements, and it should be firm and reliable. Then switch the machine operation mode to "MDI" manual data input state, input the block "T0101;" at the MDI program input position on the system panel, press the cycle start key to run the current block, and the machine tool will perform the tool change operation.

Then input the block "S500 M3;", press the cycle start key to run the current block, so that the spindle rotates at a low speed of 500r/min.

In manual mode, jog the X and Z direction control keys to move the cutter to the left side of the workpiece. Then switch to handwheel mode and move slowly with the handwheel. Before preparing to cut, the handwheel override needs to be adjusted to "×10". If the cutter moves too fast due to excessive override, the cutter and workpiece will be damaged. Control the cutter to micro-cut the end face of the blank along the X-axis direction (radial direction), and then retreat along the X-axis direction after the cutting is completed. Note that the Z-axis cannot be moved at this time.

At this moment, press the key [OFS/SET] on the system panel, press the key [刀偏] at the bottom of the display screen, and then press the key [形状] to move the cursor to the position of the corresponding cutter number, input "Z0", and press the "measure" key at the bottom of the display screen to complete the Z-direction tool setting of the cutter.

1.13.2 X-axis origin setting

Use the external turning tool to micro-turn the external surface of the workpiece, and the turning length can meet the measurement. Then retract the cutter along the Z-axis (axial), note that the X-axis cannot move at this moment. Stop the rotation of the spindle, and use a caliper or a micrometer to measure the diameter of the machined outer circle. For example, the measured outer diameter is 24.36mm.

Press the key [OFS/SET] on the system panel, press the key [刀偏] at the bottom of the display screen, and then press the key [形状] to move the cursor to the position of the corresponding cutter number, input "X24.36;", and press the "Measure" key at the bottom of the display screen to complete the X-direction tool setting for the cutter. The system interface is shown in Figure 1-17. Usually, the precision of the workpiece diameter is relatively high, so it is also necessary to input an appropriate compensation value at the corresponding "wear" position, so that the dimensional accuracy can be controlled by modifying the wear compensation value later.

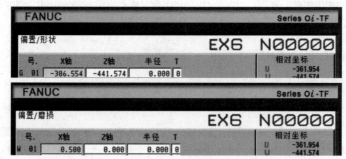

Figure 1-17 Tool compensation interface

Other cutters can be measured in a similar way. However, it should be noted that the tool holder should be moved to a safe position first, and then the cutter should be changed in "MDI" mode. When setting the tool in the Z direction, other cutters cannot be used to turn the end face of the part again, and they can only be controlled to just touch the end face that has been machined to ensure the consistency of the origin positions of different cutters. Due to the structural limitation of the cutting tool, its strength in cutting along the Z direction is poor, so the cutting tool cannot perform large-depth cutting along the axial direction, and can only perform micro-processing to avoid damage to the cutter.

1.14 Program editing

There are two operations for program entry on the machine tool system panel, one is the entry of the program number, and the other is the entry of the program statement.

Switch the operation mode to the Edit mode on the operation panel, select the button on the system panel, and then you can switch the program storage path through the "Directory" key at the bottom of the display screen , and edit the program through the "Program" key.

When entering a program, first press "Program" to enter the program editing interface.

First input the program number (range 0001~8999) , directly press the key , and then press the key and the key to change the line. At this time, it should be noted that the program number (name) and the block end character are input in two steps. If you input "O0001；" and then press the key , it will prompt "Format error" .

The input of other statements in the program is different from the program name, you can directly input the entire statement and then press the key directly. For example: .

The entered program is shown in Figure 1-18.

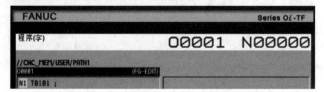

Figure 1-18　Input of CNC program

If a block needs to be deleted during the editing process, move the cursor to the position of the block to be deleted (highlighted in yellow), and then press the key to complete the deletion. If you need to modify the block selected by the cursor, you need to input the content to be changed, and press the key ![ALTER] to complete the replacement of the block content. To delete the last character in the input area ![input area icon], press the key ![CAN] to backspace.

1.15 Simulation of CNC machining program

The FANUC-0i-MF system provides the graphic verification function for CNC machining programs. Due to the difference between the system version and the factory configuration of the machine tool manufacturer, the graphic verification interface also has its own differences.

Select "AUTO" mode on the operation panel, that is, the automatic execution mode of the program, load the CNC machining program that needs graphic verification, and then press the key ![CSTM/GR] on the system panel to enter the graphic verification interface, where you can simulate the area size and viewing angle, as shown in Figure 1-19.

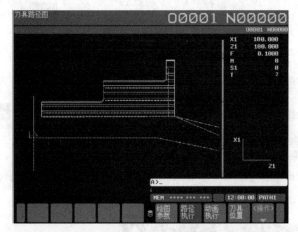

Figure 1-19　The graphic verification interface

Press the key ![MC LOCK] and key ![DRY RUN] on the operation panel, turn on the coordinate axis locking state and empty run state of the program, press the program cycle start button, and the program will start graphic verification, and the verification speed can be adjusted by adjusting the feeding override.

1.16 Part machining

After the graphic verification process has verified that there is no problem, the parts machining can be carried out. Before the parts are processed, you should understand the safety operation requirements of the machine tool in detail, and wear labor protection clothing and utensils. When processing parts, you should be familiar with the functions and positions of the operation buttons of the CNC lathe, and understand the methods of dealing with emergency situations.

Since this operation is the first piece of processing, there is a certain risk, and the operator needs to maintain a high degree of concentration throughout the process. Before automatic processing, it is necessary to adjust the rapid override to the minimum and the feed override to "0" to prevent the loss of effective control of the machine tool after startup. It is particularly important to check the correctness of the starting position of the cycle. If it is found that the cutter position does not match the coordinates, the operation should be stopped immediately to prevent danger. During the automatic operation, the processing conditions, such as surface quality and chip removal, should be carefully observed. If any abnormality occurs, manual intervention should be carried out in time.

Note: If the graphic verification operation is performed, the machine tool must be returned to the mechanical zero point after the operation is completed, and then other related operations can be performed. If the relevant operations are performed without returning to zero, it will cause coordinate offsets, and even abnormal phenomena such as collisions and program execution confusion, resulting in danger.

Select Auto mode on the operation panel, that is, the automatic execution mode of the program, load the CNC machining program that needs to be processed, and press the cycle start button to perform automatic processing. As shown in Figure 1-20.

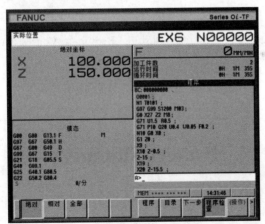

Figure 1-20　Automatic operation of the program

In this state, the program will only automatically execute the line where the cursor is located each time the cycle start button is pressed. Before starting the cycle, observe whether the distance between the cutter and the workpiece is safe. After pressing the cycle start button, control the movement speed of the machine tool through the feed override knob, in addition compare with the remaining movement amount displayed in the "Remaining Movement Amount" column of the display screen and observe the actual distance between the cutter and the workpiece at the same time. If the difference between the actual distance and the remaining movement amount is too large, the operation should be stopped and checked to avoid a collision accident. In the process of program debugging, you should also pay close attention to the "modal" status on the display screen to ensure that there is no abnormality in the spindle revolution speed, feed speed, workpiece coordinate system number, compensation status and compensation number.

1.17 Part inspection

After the parts are processed, the workpiece should be carefully cleaned, and in accordance with the relevant requirements of quality management, the processed parts should be subject to relevant inspections to ensure the production quality. Table 1-11 shows the "three-level" inspection cards for machined parts.

Table 1-11 "Three-level" inspection cards for machined parts

Part drawing number		Part name		Working step number	
Material		Inspection date		Working step name	
Inspection items	Self-inspection result	Mutual inspection result	Professional inspection		Remark
Conclusion	☐ Qualified ☐ Unqualified ☐ Repair ☐ Concession to receive Inspection signature: Date:				
Non-conforming item description					

Project Summary

As a typical machining part of CNC lathes, multi-diameter shafts are widely used in various equipment. According to equipment conditions and precision requirements, there will be some differences in the processing technology. Programmers and operators need to formulate the processing technology reasonably according to the processing conditions to improve the processing accuracy of the parts and the production efficiency.

Exercises After Class

1. Fill in the blanks

(1) CNC machine tools are mainly composed of _____ and _____.

(2) CNC machine tools are divided into _____ CNC machine tool, _____ straight cut control CNC machine tool, and _____ CNC machine tool according to the way they control the relative movement of the cutter and the workpiece.

(3) The machining action of the CNC lathe is mainly divided into two parts: the _____ and the _____.

(4) The coordinate system of the CNC machine tool adopts the _____ coordinate system.

(5) CNC programming is generally divided into _____ and _____.

(6) To shut down the CNC lathe, press _____ first and then press _____ to turn off the main power of the machine tool.

(7) The CNC lathe workpiece origin is generally set at the intersection of _____ and the right end face (or left end face), and the coordinate value of the X-axis takes the dimension _____.

2. True or false

(1) The direction of the cutter moving away from the workpiece in a certain coordinate axis is the negative direction of the coordinate axis. ( )

(2) G98 is the feed per revolution control command, and G99 is the feed per minute control command. ( )

(3) In the CNC lathe program, the use of the spindle constant surface speed controlling command G96 can improve the consistency of the surface quality of the workpiece. ( )

(4) In addition to the coordinate system setting function, the turning command G50 can also be used to limit the maximum speed of the spindle. ( )

(5) The diameter dimension programming is used when programming the CNC lathe. ( )

(6) After pressing the "Optional Stop" button on operation panel, the execution process of M01 command is the same as that of M00 command. ( )

(7) The G commands of group 00 in the FANUC system are all non-modal commands. ( )

3. Choice questions

(1) In CNC machine tools, ( ) is determined by the spindle that transmits the cutting power.

    A. X-axis coordinates   B. Z-axis coordinates
    C. Y-axis coordinates   D. C-axis coordinates

(2) The input of CNC machining program must be carried out in the ( ) mode.

    A. Manual   B. Auto   C. MDI   D. Edit

(3) three-jaw self centering chucks on lathes and gad tongs on milling machines belong to ( ).

    A. General fixture   B. Special fixture
    C. Combined fixture   D. Follow fixture

(4) The movement speed of commanded G00 is designated by (   ).

A. Machine tool parameters   B. CNC program

C. Operation panel   D. Random

(5) Among the following trajectories, the path of G00 may be (   ).

A. straight line   B. oblique line

C. broken line   D. all of the above are possible

(6) When determining the coordinate axis of the CNC machine tool, (   ) should generally be determined first.

A. X-axis   B. Y-axis   C. Z-axis   D. A-axis

(7) The buttons "F0", "F25", "F50" and "F100" are used to control the (   ) override of the CNC machine tool.

A. manual feed   B. automatic feed

C. incremental feed   D. rapid feed

(8) After executing the block "G00 X20.0 Z30.0; G01 X10.0 W20.0 F0.2; U-40.0 W-70.0;", the cutter position of the workpiece coordinate system is (   ).

A. (X-40.0,Z-70.0)   B. (X-30.0,Z-50.0)

C. (X-30.0,Z-20.0)   D. (X-10.0,Z-20.0)

4. Short answer questions

(1) Briefly describe the types of the fixtures of CNC lathe.

(2) Briefly describe selection principle of cutter blade material.

(3) Briefly describe what should be paid attention to when the vertical rough machining cycle instruction G71 is used for turning the outer circle.

5. Comprehensive programming questions

Determine the processing technology according to the part drawing (Figure 1-21), write the program, and automatically process it.

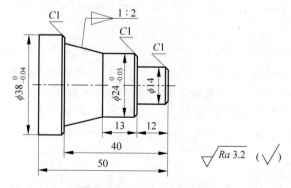

Figure 1-21   Part drawing

Self-learning test score sheet is shown in Table 1-12.

Table 1-12 Self-learning Test Score Sheet

Tasks	Task requirements	Score	Scoring rules	Score	Remark
Learn key knowledge points	(1) Know the classification of CNC lathes and the structure and main parameters of CK6150e (2) Understand the characteristics of turning (3) Familiar with the classification of commonly used tools and can make the correct selection of tools (4) Know the general fixtures and measuring tools of CNC lathes (5) Master the basic format of CNC turning program and basic preparation function instructions	20	Understand and master		
Technological preparation	(1) Be able to correctly read the shaft part drawings (2) Can determine the processing technology independently and fill in the technology documents correctly (3) Be able to write the correct processing program according to the processing process	30	Understand and master		
Hands-on training	(1) The corresponding equipment and utensils will be selected correctly (2) Master the selection and installation method of outer circle turning tool and cut-off tools (3) Can correctly operate the CNC lathe and adjust the processing parameters according to the machining situation	50	(1) Understand and master (2) Operation process		

Ideological and Political Classroom

Project 2 Programming and Machining Training for Handle Turning

➢ Mind map

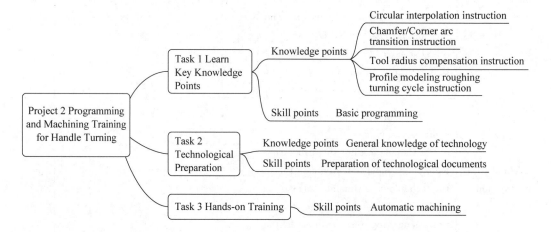

➢ Learning objectives

Knowledge objectives

(1) Know the machining characteristics of curved contours.

(2) Understand the meaning of each parameter of the profile modeling roughing turning cycle instruction.

Ability objectives

(1) Be able to independently determine the process routine and fill in the technological documents correctly.

(2) Be able to operate the CNC lathe correctly and adjust the machining parameters according to the machining conditions.

(3) Be able to select measuring tools reasonably according to the accuracy and the structural characteristics of parts, and be able to measure the relevant dimensions correctly and normatively.

Literacy goals

(1) Develop students'scientific spirit and attitude.

(2) Cultivate students'engineering awareness.

(3) Develop students'teamwork skills.

Task introduction

A handle is mainly used for hand-held operation by the staff, and usually consists of a connecting part and a hand-held part. The connecting part is used to connect with other parts and transmit motion, and the hand-held part is used for manual operation by the staff. Because the hand-held part is in direct contact with the hand during the working process, the surface is usually required to be smooth and free of burrs to prevent unnecessary personal injury. Handles are widely used in production and life as manually operated parts.

According to the requirements of the handle part drawing Figure 2-1, the processing technology is formulated, the CNC machining program is developed, and the processing of handle part is completed. The blank material of this part is 45 steel, which is quenched and tempered and requires surface smoothing.

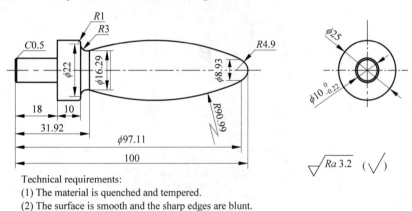

Technical requirements:
(1) The material is quenched and tempered.
(2) The surface is smooth and the sharp edges are blunt.

Figure 2-1 The part drawing of handle

Task 1 Learn Key Knowledge Points

2.1 Basic instructions

2.1.1 Circular interpolation instruction

The circular interpolation instruction is to move the cutter in the form of a arc from its current position to the position required by the instruction at the speed rate specified by the F code. Its characteristic is that each coordinate moves in a linkage manner and moves in an arbitrary arc according to the feed rate F. According to the circular motion direction, it is divided into clockwise circular interpolation and counterclockwise circular interpolation, and the corresponding instructions are G02 and G03 instruction respectively. When programming, the final position is dedicated by X and Z (or U and W). The arc radius is specified by R ___ (when the central angle is less than or equal to 180°, R is a positive value, and the center angle is greater than 180° and less than 360°, R is a negative value), or use I __ K __ to specify the coordinate increment of the arc center relative to the arc start point, which can be omitted when *I* and *K* are "0".

Instruction format:

G02 X __ Z __ R __ F __ ; (clockwise circular interpolation)
G03 X __ Z __ R __ F __ ; (counter clockwise circular interpolation).

or

G02 X __ Z __ I __ K __ F __ ;
G02 X __ Z __ I __ K __ F __ ;

As shown in Figure 2-2, the cutter moves counterclockwise from point *A* to point *B* at the specified speed *F*, and then moves clockwise from point *B* to point *C* at the specified speed *F*. Its corresponding program is shown in Table 2-1.

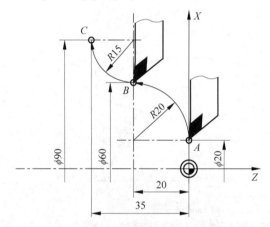

Figure 2-2 A path for circular interpolation instruction

Programming with circular interpolation instruction is shown in Table 2-1。

Table 2-1 Programming with circular interpolation instruction

Programming with radius *R*	Programming with center vectors *I* and *K*
...	...
G03 X60 Z-20 R20 F0.2;	G03 X60 Z-20 I0 K-20 F0.2;
G02 X90 Z-35 R15;	G02 X90 Z-35 AND15 K0;
...	...

2.1.2 Chamfer/Corner arc transition instruction

When turning parts, sharp edges usually need to be chamfered or rounded to facilitate post-assembly and meet safety requirements. Programming can be simplified by using chamfer transition instructions.

1. Chamfer transition instruction C __

It is used for inserting the chamfering between two straight profiles.

Instruction format:

```
G01 X__ Z__ C__ F__;
```

For example, when programming the outer circle contour as shown in Figure 2-3, the intersection point D can be programmed directly without programming the straight contour segment AB. Its corresponding program is shown in Table 2-2.

2. Corner arc transition R __

It is used for inserting a transitional arc between two straight profiles.

Instruction format:

```
G01 X__ Z__ R__ F__;
```

For example, when programming the outer circle contour as shown in Figure 2-4, the intersection point D can be programmed directly without programming the arc contour segment AB. Its corresponding program is shown in Table 2-3.

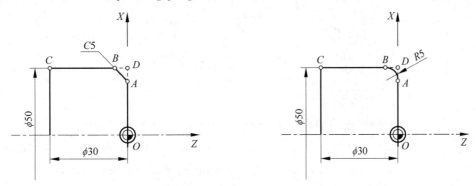

Figure 2-3 A path for chamfer transition instruction   Figure 2-4 A path for corner arc transition instruction

Programming with chamfer transition instruction C is shown in Table 2-2.

Table 2-2 Programming with chamfer transition instruction C

Programming with the linear interpolation instruction G01	Programming with chamfer transition instruction C
...	...
G01 X40;	G01 X50 C5;
X50 Z-5;	Z-30;
Z-30;	...
...	...

Programming with corner arc transition instruction R is shown in Table 2-3.

Table 2-3 Programming with corner arc transition instruction R

Programming with linear interpolation instruction G01	Programming with corner arc transition instruction R
...	...
G01 X40 F0.1;	G01 X50 R5 F0.1;
X50 Z-5;	Z-30;
Z-30;	...
...	...

Note: Programming with chamfer/corner arc transition instruction is only applicable between two intersecting straight lines, and the actual length of the straight lines cannot be "0". The coordinate of the programmed end point is that of the intersection point.

2.1.3 Tool radius compensation instructions

For CNC turning, in order to improve the tool durability, the tool nose is usually made into an arc shape. As shown in Figure 2-5, if the tool radius compensation is not established, when the arc tool nose is used to machine other contours except the end face and the cylindrical surface, it will cause overcut or undercut. In order to solve this problem, various CNC systems have introduced the nose radius compensation instructions.

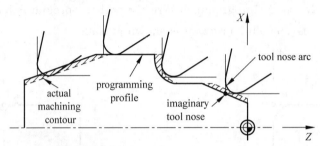

Figure 2-5 Influence of nose radius compensation on machining

The arc positions of different tools are not the same, and the compensation directions are also different. In order to distinguish various tools, the tool nose radius position code is introduced, as shown in Figure 2-6. When establishing and canceling nose radius compensation, it needs to be used in conjunction with the linear interpolation instruction G01 or the rapid positioning instruction G00. The compensation can be divided into three processes. Turn on tool radius compensation: done before the cutter enters the part contour; Execute tool radius compensation: the cutter continuously processes on the part contour; Turn off tool radius compensation: Implemented after the cutter has left the contour.

(1) Turn on tool radius compensation left G41, shown in Figure 2-7, Instruction format:

G01 (G00) G41 X __ Z __ F __ ;

(2) Turn on tool radius compensation right G42, shown in Figure 2-7, Instruction format:

G01 (G00) G42 X __ Z __ F __ ;

(3) Turn off tool radius compensation G40, Instruction format:

G01 (G00) G40 X __ Z __ F __ ;

In the machine tool, it is also necessary to input the corresponding tool nose position code in the tool offset parameter, as shown in Figure 2-8, for example, the code of the external turning tool is 3, and it is input into the corresponding position.

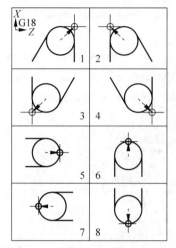

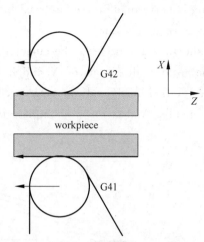

Figure 2-6 Tool radius position code   Figure 2-7 Tool radius compensation direction

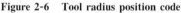

Figure 2-8 Tool nose position code

2.1.4 Profile modeling roughing turning cycle instruction G73

Instruction format:

G73 U (Δi)W(Δk) R (d);
G73 P (ns) Q (nf) U (Δu) W (Δw) F(f);

Instruction description:

Δi represents the retraction distance along the X-axis (radial), that is, the total allowance to be machined in this direction.

Δk represents the retraction distance along the Z-axis (axial), that is, the total allowance to be machined in this direction.

Δd represents the number of rough machining.

ns indicates the starting block number in the finishing contour block.

nf indicates the end block number in the finishing contour block.

Δu represents the finishing allowance in the X direction, and is a negative value when machining inner contours.

Δw represents the finishing allowance in the Z direction, and the value is generally 0.05~0.1mm.

The G73 command is mainly used for the contour cutting of castings, forgings, and non-monotonic changes in radial dimensions. For the case where the blank is very close to the contour of the part, the processing efficiency is high. However, when multi-tool machining is performed on a cylindrical blank, there are many empty tool paths, and the efficiency is relatively low.

As shown in Figure 2-9, the cutter first moves rapidly from the cycle starting point along the negative directions of X and Z axes by a distance of $\Delta i + \Delta u/2$ and $\Delta k + \Delta w$ respectively, then feeds and machines according to the contour path, finally retreats and returns quickly. And so on, to complete the roughing.

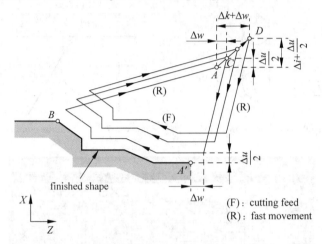

Figure 2-9　Profile modeling roughing turning cycle instruction

Precautions:

(1) When using G73 instruction for roughing, the F and S commands included in the ns~nf block are invalid for the rough turning cycle.

(2) Subprograms cannot be called in the program segment whose sequence number is ns~nf.

Task 2　Technological Preparation

2.2　Part drawing analysis

According to the use requirements of the parts, 45 steel is selected as the blank material of handle part, and the blanking dimension is set as $\phi 30 \times 105$. Taking the outer circle of $\phi 30$ as the rough reference, rough and finish the right part of the handle to the required size. Then turn it around and clamp the outer circle of $\phi 25$ (pay attention to protecting the surface when clamping), and process the left end face of the multi-diameter shaft with diameters of $\phi 10$ and $\phi 25$ and the outer circle to the required size.

Note that when turning the outer circle of $\phi 25$, the turning length should be sufficient. In addition, when clamping the blank, attention should be paid to the extended length of the bar to avoid collision between the cutter and the chuck.

2.3 Technological Design

According to the analysis of the part drawing, the technological process is designed as shown in Table 2-4.

Table 2-4 Technological process card

Machining process card	Product model	CLJG-01		Part number	HS-01	Page 1	
	Product name	Geneva mechanism		Part name	Handle	Total 1 page	
Material grade	C45	Blank size	$\phi 30 \times 105$	Blank quanlity	kg	Quantity	1
Working Procedure				Work Section	Technical equipment	Man-hour/min	
No.	Name	Content				Preparation & Conclusion	Single Piece
5	Preparation	Prepare the material according to the size of $\phi 25 \times 60$.		Outsourcing	Sawing machine		
10	Turning	Taking the excircle of $\phi 30$ as the rough reference, rough and finish the handle part		Turning	Lathe, outside micrometer	45	30
15	Turning	Taking the processed handle and $\phi 25$ outer circle as the fine reference (as far as possible coaxial), rough and finish the surface and end face of the outer circle of $\phi 10$ and $\phi 25$, and ensure the length requirements		Turning	Lathe, outside micrometer	45	30
20	Cleaning	Clean the workpiece, debur sharp corner		Locksmith			5
25	Inspection	Check the workpiece dimensions		Examination			5

Based on the 10th working procedure of turning, this training task is designed, and the corresponding working procedure card is formulated as shown in Table 2-5.

Table 2-5 Working procedure card for turning

Machining working procedure card	Product model	CLJG-01	Part number	HS-01	Page 1
	Product name	Geneva mechanism	Part name	Handle	Total 1 page

Procedure No.		10,15
Procedure name		Turning
Material		C45
Equipment		CNC lathe
Equipment model		CK6150e
Jig		three-jaw self-centering chuck
Measuring tool		Vernier caliper
		micrometer
Preparation & conclusion time		90min
Single-piece time		60min

Technical requirements:
(1) The material is quenched and tempered.
(2) The surface is smooth and the sharp edges are blunt.

Continued

Work steps	Content	Cutters	S/(r/min)	F/(mm/r)	a_p/mm	Step hours/min	
						Mechanical	Auxiliary
1	Workpiece installation						5
2	The outer surface and end face of ϕ10 and ϕ20 are roughed, and the finishing allowance is 0.2mm	Outer circle roughing turning tool	1200	0.2	1.5	15	
3	Finishing the outer surface and end face of ϕ10 and ϕ20	Outer circle finishing turning tool	1500	0.1	0.2	10	
4	Dismantling and cleaning workpieces						5

2.4 CNC machining programming

According to the technology of the working procedure, the machining program is written as shown in Table 2-6.

Table 2-6 CNC program for handle turning

No.	Program Statement	Annotation
	O0001	**CNC machining program for the right part**
N1	T0101;	Call the outer circle turning tool
	G97 G99 S1000 M03;	Set constant rotative speed control, unit of feed mount is mm/r, spindle speed is 1000r/min, forward rotation
	G0 X32 Z2 M8;	Rapidly locate to the beginning of the cycle (X32, Z2) and turn on the coolant
	G73 U15 W0 R10;	Call the profile modeling roughing turning cycle instruction
	G73 P10 Q20 U0.4 W0 F0.16;	The radial finishing allowance is 0.4mm (0.2mm per side), the axial finishing allowance is 0.05mm, the feed rate is 0.2mm/r
N10	G0 X0;	
	G1 G42 Z0;	
	G3 X8.93 Z-2.89 R4.9;	
	G3 X16.29 Z-68.08 R90.99;	
	G2 X22 Z-72 R3;	
	G1 X25 R1;	
	Z-83;	
N20	G1 G40 X32;	
	G0 X100 Z150;	Quickly retract to (X100, Z150)
	M5;	Stop the spindle
	M9;	Turn off the coolant
	M01;	Optional stop (the optional stop button needs to be pushed) to observe the completion of roughing

Continued

No.	Program Statement	Annotation
N2	T00101;	Call the outer circle turning tool
	G97 G99 S1200 M3;	Set constant speed control, feed unit m/r, spindle 1200r/min, forward rotation
	G0 X32 Z2 M8;	Quickly locate to the start of the cycle (X32, Z2) and turn on the coolant
	G70 P10 Q20 F0.1;	Invoke the finishing cycle instruction with a feed of 0.1mm/r
	G0 X100 Z150;	Quickly retract to (X100, Z150)
	M5;	Stop spindle
	M9;	Turn off the coolant
	M30;	The program ends
	O0002;	**CNC machining program for the left part**
N3	T0101;	Call the outer circle turning tool
	G97 G99 S1000 M3;	Set constant rotative speed control, unit of feed mount is mm/r, spindle speed is 1000r/min, forward rotation
	G0 X32 Z10;	Quickly locate to (X32, Z10)
	G71 U1.5 R0.5;	Call the longitudinal roughing cycle instruction, the cutting depth is 1.5mm, and the retraction amount is 0.5mm
	G71 P30 Q40 U0.4 W0.05 F0.18;	The radial finishing allowance is 0.4mm (0.2mm on one side), and the axial finishing allowance is 0.05mm, and the feed rate is 0.18mm/r
N30	G0 X0;	
	G1 Z0;	
	X10 C0.5;	
	Z-18;	
N40	X27;	
	G0 X100;	Radial retract
	Z150;	Axial retract
	M5;	Stop the spindle
	M9;	Turn off the coolant
	M01;	Optional stop (the optional stop button needs to be pushed) to observe the completion of roughing
N4	T00101;	Call the outer circle turning tool
	G97 G99 S1200 M3;	Set constant rotative speed control, unit of feed mount is mm/r, spindle speed is 1200r/min, forward rotation
	G0 X32 Z2 M8;	Quickly locate to the start of the cycle (X32, Z2) and turn on the coolant
	G70 P30 Q40 F0.1;	Call the finishing cycle instruction with a feed of 0.1mm/r
	G0 X100 Z150;	Quickly retract to (X100, Z150)
	M5;	Stop the spindle
	M9;	Turn off the coolant
	M30;	End of the program

Task 3　Hands-on Training

2.5　Equipment and appliances

Equipment: CK6150e CNC lathe.
Cutters: the outer circle turning tool, the cut-off tool with a blade width of 3mm.
Fixture: the three-jaw self-centering chuck.
Tools: chuck wrenches, tool holder wrenches, etc.
Gauges: the 0~150mm vernier caliper and the 0~25mm outside micrometer.
Blank: $\phi 30 \times 105$.
Auxiliary appliances: cutter shim, brushes and so on.

2.6　Check before powering on

Refer to Table 2-7 to check the machine status.

Table 2-7　Preparing card for machine start-up

	Check item	Test result	Abnormal description
Mechanical part	Spindle		
	Feed part		
	Tool holder		
	three-jaw self centering chuck		
Electrical part	Main power supply		
	Cooling fan		
CNC system	Electrical components		
	Controlling part		
	Driving section		
Auxiliary part	Cooling system		
	Compressed air system		
	Lubricating system		

2.7　Preparation before machining

Before machining, the tools required for this task should be prepared and installed correctly. The origin of the workpiece is set according to the process requirements, and the CNC machining program is entered and verified.

2.8 Part machining

After the graphic verification process has verified that there is no problem, the parts machining can be carried out. Before the parts are processed, you should understand the safety operation requirements of the machine tool in detail, and wear labor protection clothing and utensils. When processing parts, you should be familiar with the functions and positions of the operation buttons of the CNC lathe, and understand the methods of dealing with emergency situations. During the machining process, especially before cutting, the actual distance between the cutter and the workpiece should be observed, based on the remaining movement amount displayed in the "Remaining Movement Amount" column on the display screen is referenced. When the difference between the actual distance and the remaining movement amount is too large, the vehicle should be stopped and checked to avoid collision. If there is any abnormality, the machine tool movement should be stopped in time.

2.9 Part inspection

After the parts are processed, the workpiece should be carefully cleaned, and in accordance with the relevant requirements of quality management, the processed parts should be subject to relevant inspections to ensure the production quality. Table 2-8 shows the "three-level" inspection cards for machined parts.

Table 2-8 "Three-level" inspection cards for machined parts

Part drawing number		Part name		Working step number	
Material		Inspection date		Working step name	
Inspection items	Self-inspection result	Mutual inspection result	Professional inspection	Remark	
Conclusion	☐ Qualified　☐ Unqualified　☐ Repair　☐ Concession to receive 　　　　　　　　　　　　　　　　　　Inspection signature: 　　　　　　　　　　　　　　　　　　Date:				
Non-conforming item description					

Project Summary

As a typical machining part of CNC lathes, the handle is widely used in various equipment. According to equipment conditions and precision requirements, there will be some differences in the processing technology. Programmers and operators need to formulate the processing technology reasonably according to the processing conditions to improve the processing accuracy of the parts and the production efficiency.

Exercises After Class

1. Fill in the blanks

(1) The process of tool radius compensation is divided into _____, _____, and _____.

(2) There are two types of tool radius compensation for CNC lathes: _____ and _____.

(3) The main processing methods for the outer surface are _____ and _____.

(4) The instruction to turn off the tool radius compensation is _____.

(5) The machining contour of the profile modeling rough turning cycle instruction G73 is written between _____ and _____.

2. True or false

(1) When the circular interpolation is programmed with R code, if the central angle of the circular arc is less than 180°, R takes a negative value.   (   )

(2) In the FANUC system, the establishment and cancellation of the tool radius compensation mode are only valid in the instruction G00 and G01 movement command modes.   (   )

(3) In the FANUC system, if non-monotonic contours are programmed in the ns~nf block in the G71 instruction, an alarm will be generated during the execution of G71.   (   )

(4) Viewed from the negative direction of the Y-axis, the clockwise circular interpolation instruction is expressed by G02.   (   )

(5) The chamfer transition instruction of FANUC system can be used for chamfering two intersecting straight lines at any angle.   (   )

(6) After the G70 cycle ends, the cutter returns to the reference point.   (   )

3. Choice questions

(1) The *I* and *K* values in arc programming refer to the vector values of (   ).

  A. the starting point to the center of the circle

  B. the end point to the center of the circle

C. the center of the circle to the starting point

  D. the center of the circle to the end point

(2) The selection of cutting amount for finishing is generally based on (　　).

  A. Improve productivity　　B. Reduce cutting power

  C. Guarantee the processing quality　D. Improve the surface quality

(3) When the inner and outer circle compound canned cycle is used for programming, if the (　　) instruction is included in the blocks ns~nf, the program alarm will not be generated during the program execution.

  A. canned cycle

  B. reference point return

  C. compound canned cycle

  D. arc machining with a central angle greater than 90°

(4) In the following canned cycles, the canned cycle that the block with sequence number ns must feed in the Z direction and the X coordinate cannot appear is (　　).

  A. G71　　B. G72　　C. G73　　D. G74

(5) It is more appropriate to use the (　　) instruction as the roughing process cycle instruction for the workpieces of casting molding and rough turning.

  A. G71　　B. G72　　C. G73　　D. G74

(6) When using the G02/G03 instruction, the incorrect description about programming in corner arc transition instruction is (　　).

  A. Full circle machining cannot be programmed in this way.

  B. This way has the same effect as using I, J, and K.

  C. For the arc whose central angle is greater than 180°, R takes a positive value.

  D. R can take positive or negative values, but the processing trajectory is different.

4. Short answer questions

(1) Briefly describe how to determine the direction of the tool radius compensation.

(2) Briefly describe the functions of the tool radius compensation.

(3) Briefly describe the main occasions of profile modeling roughing turning cycle instruction G73 used.

Self-learning test score sheet is shown in Table 2-9.

Table 2-9 Self-learning Test Score Sheet

Task	Task requirement	Score	Scoring rule	Score	Remark
Learn key knowledge points	(1) Master the use of chamfer transition instruction (2) Master the use of the tool radius compensation instruction (3) Understand the meaning of the parameters of the profile modeling roughing cycle instruction	20	Understand and master		
Technological preparation	(1) Be able to read part drawings correctly (2) Be able to independently determine the processing route and fill in the process documents correctly (3) Be able to write the correct processing program according to the machining process	30	Understand and master		
Hands-on training	(1) Be able to select measuring tools reasonably according to the accuracy and the structural characteristics of parts, and be able to measure the relevant dimensions correctly and normatively (2) Master the operation process of handle turning (3) Be able to operate the CNC lathe correctly and adjust the processing parameters according to the processing conditions	50	(1) Understand and master (2) Operation process		

Ideological and Political Classroom

Project 3 Programming and Machining Training for Drive Shaft Turning

> Mind map

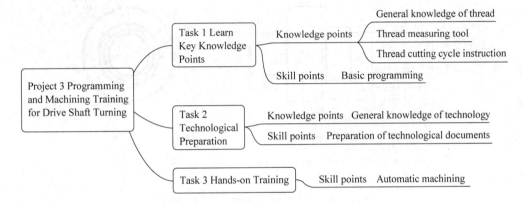

> Learning Objectives

Knowledge Objectives

(1) Know the machining characteristics of threaded shafts.

(2) Understand the meaning of each parameter of the thread cutting cycle instruction.

Competency Objectives

(1) Be able to independently determine the process routine and fill in the technological documents correctly.

(2) Be able to operate the CNC lathe correctly and adjust the machining parameters according to the machining conditions.

(3) Be able to select measuring tools reasonably according to the accuracy and the structural characteristics of parts, and be able to measure the relevant dimensions correctly and normatively.

Literacy Goals

(1) Develop students' scientific spirit and attitude.

(2) Cultivate students' engineering awareness.

(3) Develop students' teamwork skills.

➢ **Task Introduction**

The threaded shaft is widely used in production and life, and is mainly composed of a multi-diameter shaft and the threaded part. The external thread part is used to connect with the nut and plays a fixed role.

According to the requirements of the drive shaft part drawing (shown in Figure 3-1), the processing technology is formulated, the CNC machining program is developed, and the processing of the drive shaft is completed. The blank material of this part is 45 steel, which is quenched and tempered and requires surface smoothing.

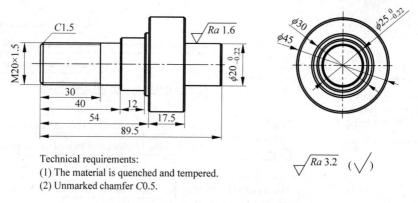

Figure 3-1 The part drawing of drive shaft

Task 1   Learn Key Knowledge Points

3.1 General knowledge of thread

3.1.1 Classification of threads

According to the application, screw threads can be divided into connecting threads (fastening threads), transmission threads, pipe threads and special threads. According to the thread form, screw threads can be divided into triangular threads, trapezoidal threads, rectangular threads, sawtooth threads and arc threads. According to the revolving direction of screw thread, threads can be divided into left-hand threads and right-hand threads. According to the number of spiral lines, threads can be divided into single-start threads and multi-start threads. According to the shape of the thread body which is a cylinder or cone where the screw thread is made, screw threads are divided into parallel threads and tapered threads.

3.1.2 Representation of metric threads

The ISO standard metric thread angle is 60 degrees, which can be divided into two types: coarse pitch thread and fine pitch thread. The dimension representation and meaning of each part are shown in Figure 3-2.

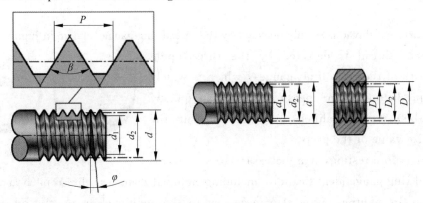

Figure 3-2 Dimension representation of metric threads

The meaning of each parameter is:

P: thread pitch.

β: thread angle.

d_1: minor diameter of external thread.

D_1: minor diameter of internal thread.

d_2: pitch diameter of external thread.

D_2: pitch diameter of internal thread.

d: major diameter of external thread.

D: major diameter of internal thread.

φ: lead angle of the thread.

The marking of the thread is mainly composed of thread code, nominal diameter $d(D)$, pitch P, tolerance zone code and length of thread engagement. The common metric thread code is represented by the letter "M". The common thread is usually right-hand thread, no special marking is required. If it is left-hand thread, it needs to be marked with "LH". If the pitch, P, is not marked, it is indicated as a coarse pitch thread, and if it is a fine pitch thread, the pitch needs to be marked.

For example, M24×1.5-6g represents a common metric fine-pitch thread with a nominal diameter of 24mm, pitch of 1.5mm and tolerance zone code of 6g.

3.1.3 Calculation of dimensions related to thread turning

Major diameter of external thread d = nominal diameter. In actual processing, since the tolerance of external thread is mainly lower deviation, the major diameter of external thread is usually processed to $d-0.1P$, and the small diameter of external thread is $d=D-1.3P$.

3.2 Thread measuring tool

The external thread is mainly detected by the thread ring gauge (shown in Figure 3-3), and the internal thread is detected by the thread plug gauge. Both of them are limit gauges, which is widely used in production. The measuring tool can quickly detect if the parts are qualified, but cannot measure the specific size value of the parts.

Figure 3-3　Thread ring gauge

Note: Before testing, it is necessary to see clearly the thread ring gauge identification, including nominal diameter, thread pitch and tolerance zone code. In addition, it is also necessary to thoroughly clean up the oil stains and impurities of the tested thread, and then after the ring gauge is aligned with the tested thread, rotate the ring gauge with the thumb and forefinger to make it screw together in a free state. Passing the whole length of the thread is judged as qualified, otherwise it is judged as unqualified.

3.3 Basic instruction

Thread cutting cycle instruction G92 is mainly used for processing parallel and tapered threads with equal pitch. Instruction format:

G92 X(U)__ Z(W)__ R__ F__;

Instruction description:

X(U)__ Z(W)__ indicates the coordinate of the thread end point.

R__ represents the coordinate increment of the start and end points of the tapered thread in the X-axis direction (radius value). R is "0" in the parallel thread cutting cycle, which can be omitted.

F__ represents the lead L of the screw thread. For a single-start thread, lead equals pitch P.

As shown in Figure 3-4, the cutter first feeds rapidly along the X-axis from the starting point of the cycle, then processes according to the programmed path, and then retracts and returns quickly.

Precautions:

(1) During the execution of the thread cutting cycle instruction, the spindle and feed override are invalid.

(2) If the feed pause is applied during thread cutting (the second action), the cutter immediately retracts while performing chamfering, and returns in the order of the second axis (X-axis) and the first axis (Z-axis) of the plane to the starting point.

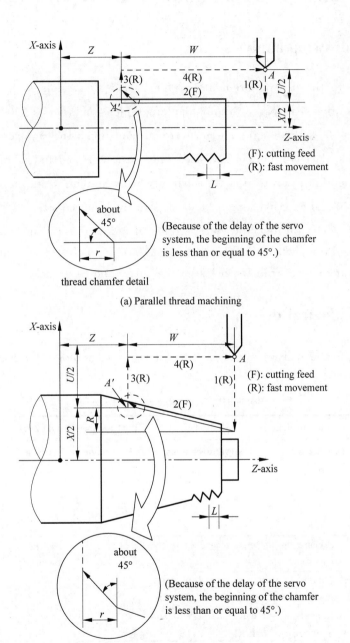

Figure 3-4　Thread cutting cycle instruction

Task 2 Technological Preparation

3.4 Part drawing analysis

According to the use requirements of the parts, 45 steel is selected as the blank material of the drive shaft part, and the blanking dimension is set as $\phi 50 \times 95$. Taking the outer circle of $\phi 50$ as the rough reference, rough and finish the right end face, $\phi 20$ and $\phi 45$ cylindrical surfaces to the required size. Then turn it around and clamp the outer circle of $\phi 20$ (pay attention to protecting the surface when clamping), and process the left end face of $\phi 25$, $\phi 30$ and external thread $M20 \times 1.5$ to the required size.

Note that when turning the outer circle of $\phi 25$, the turning length should be sufficient. In addition, when clamping the blank, attention should be paid to the extended length of the bar to avoid collision between the cutter and the chuck.

3.5 Technological design

According to the analysis of the part drawing, the technological process is designed as shown in Table 3-1.

Table 3-1 Technological process card

Machining process card	Product model	CLJG-01		Part number	QD Z-01	Page 1	
	Product name	Geneva mechanism		Part name	Drive shaft	Total 1 page	
Material grade	C45	Blank size	$\phi 50 \times 95$	Blank quanlity	kg	Quantity	1
Working Procedure				Work Section	Technical equipment	Man-hour/min	
No.	Name	Content				Preparation & Conclusion	Single Piece
5	Preparation	Prepare the material according to the size of $\phi 50 \times 95$		Outsourcing	Sawing machine		
10	Turning	Taking the outer circle of $\phi 50$ as the rough reference, rough and finish the right end face of drive shaft, $\phi 20$ and $\phi 45$ cylindrical surfaces		Turning	Lathe, outside micrometer	45	30
15	Turning	Taking the processed part of drive shaft and $\phi 20$ outer circle as the fine reference (as far as possible coaxial), rough and finish the outer circle of $\phi 25$ and $\phi 30$ and external thread $M20 \times 1.5$, and ensure the length requirements		Turning	Lathe, outside micrometer	45	30
20	Cleaning	Clean the workpiece, debur sharp corner		Locksmith			5
25	Inspection	Check the workpiece dimensions		Examination			5

The working procedures of this training task is designed based on the 10th and 15th working procedures, and the corresponding working procedure card is formulated as shown in Table 3-2.

Table 3-2 Working procedure card for turning

Machining working procedure card	Product model	CLJG-01	Part number	QD Z-01	Page 1
	Product name	Geneva mechanism	Part name	Drive shaft	Total 1 page

	Procedure No.	10, 15
	Procedure name	Turning
	Material	C45
	Equipment	CNC lathe
	Equipment model	CK6150e
	Fixture	Three-jaw self-centering chuck
	Measuring tool	Vernier caliper
		Micrometer
	Preparation & conclusion time	90min
	Single-piece time	60min

Technical requirements:
(1) Material is quenched and tempered.
(2) Unmarked chamfer C0.5.

Work steps	Content	Cutters	S/ (r/min)	F/ (mm/r)	a_p/ mm	Step hours/min	
						Mechanical	Auxiliary
1	Workpiece installation						5
2	Roughing ϕ20 and ϕ45 outer surface and end face, and finishing allowance is 0.2mm	Outer circle roughing turning tool	1200	0.2	1.5	15	
3	Finishing ϕ20 and ϕ45 outer surface and end face	Outer circle finishing turning tool	1500	0.1	0.2	10	
4	Roughing ϕ20, ϕ25 and ϕ30 outer surface and end face, and finishing allowance is 0.2mm	Cutting tool with a width of 3mm	700	0.07	0.1	15	
5	Machining M20×1.5 external thread	External thread turning tool	500	1.5		10	5
6	Dismantling and cleaning workpieces						5

3.6 CNC machining programming

According to the technology of the working procedure, the right end and left end processing programs are written separately, as shown in Table 3-3 and Table 3-4.

Table 3-3 CNC machining program for the right end of the drive shaft

No.	Program statement	Annotation
	O0001;	CNC machining program for the right end of drive shaft
N1	T0101;	Call the outer circle turning tool
	G97 G99 S800 M03;	Set constant rotative speed control, unit of feed mount is mm/r, spindle speed is 800r/min, forward rotation
	G0 X52 Z2 M8;	Rapidly locate to the beginning of the cycle (X52, Z2) and turn on the coolant
	G71 U1.5 R0.5;	Call the longitudinal roughing turning cycle instruction
	G71 P10 Q20 U0.4 W0.2;	The radial finishing allowance is 0.4mm (0.2mm per side), the axial finishing allowance is 0.05mm, the feed rate is 0.2mm/r
N10	G0 X0;	
	G1 G42 Z0;	
	X20 C0.5;	
	Z-18;	
	X45 C0.5;	
	Z-40;	
N20	G1 G40 X52;	
	G0 X100 Z150;	Quickly retract to (X100, Z150)
	M5;	Stop spindle
	M9;	Turn off the coolant
	M01;	Optional pause (the selective pause button needs to be pressed to work) to observe the completion of roughing
N2	T00101;	Call the outer circle turning tool
	G97 G99 S900 M3;	Set constant rotative speed control, unit of feed mount is mm/r, spindle speed is 900r/min, forward rotation
	G0 X52 Z2 M8;	Rapidly locate to the beginning of the cycle (X32, Z2) and turn on the coolant
	G70 P10 Q20 F0.1;	Call the finishing cycle instruction with a feed of 0.1mm/r
	G0 X100 Z150;	Quickly retract to (X100, Z150)
	M5;	Stop spindle
	M9;	Turn off the coolant
	M30;	The end of the program

Table 3-4 CNC machining program for the left end of the drive shaft

No.	Program statement	Annotation
	O0002;	CNC machining program for the left end of drive shaft
N1	T0101;	Call the outer circle turning tool
	G97 G99 S900 M3;	Set constant rotative speed control, unit of feed mount is mm/r, spindle speed is 900r/min, forward rotation
	G0 X52 Z10;	Rapidly locate to the beginning of the cycle (X52, Z10) and turn on the coolant
	G71 U1.5 R0.5;	Call the longitudinal roughing turning cycle instruction, the cutting depth is 1.5mm, and the retraction amount is 0.5mm
	G71 P30 Q40 U0.4 W0.05 F0.18;	The radial finishing allowance is 0.4mm (0.2mm per side), the axial finishing allowance is 0.05mm, the feed rate is 0.18mm/r

Continued

No.	Program statement	Annotation
N30	G0 X0;	
	G1 G42 Z0;	
	X19.85 C1.5;	
	Z-40;	
	X30 C0.5;	
	Z-52;	
	X35 C0.5;	
	Z-54;	
	X44;	
	X46 Z-55;	
N40	G1 G40 X52;	
	G0 X100;	Radial retract
	Z150;	Axial retract
	M5;	Stop the spindle
	M9;	Turn off the coolant
	M01;	Optional stop (the optional stop button needs to be pushed) to observe the completion of roughing
N2	T00101;	Call the outer circle turning tool
	G97 G99 S900 M3;	Set constant rotative speed control, unit of feed mount is mm/r, spindle speed is 900r/min, forward rotation
	G0 X52 Z2 M8;	Quickly locate to the start of the cycle ($X52$, $Z2$) and turn on the coolant
	G70 P30 Q40 F0.1;	Call the finishing cycle instruction with a feed of 0.1mm/r
	G0 X100 Z150;	Quickly retract to ($X100$, $Z150$)
	M5;	Stop the spindle
	M9;	Turn off the coolant
	M01;	Optional stop (the optional stop button needs to be pushed) to observe the completion of roughing
N3	T0202;	Call the external thread turning tool
	G97 G99 S500 M3;	Set constant rotative speed control, unit of feed mount is mm/r, spindle speed is 500r/min, forward rotation
	G0 X22 Z5 M8;	Quickly locate to the start of the cycle ($X22$, $Z2$) and turn on the coolant
	G92 X19 Z-30 F1.5;	Call the thread cutting cycle instruction G92 to perform the first cutting, and the feed rate F is 1.5mm/min
	X18.4;	The second cutting
	X18.2;	The third cutting
	X18.05;	The fourth cutting
	X18.05;	Smooth thread form surface
	G0 X100 Z150;	Quickly retract to ($X100$, $Z150$)
	M5;	Stop the spindle
	M9;	Turn off the coolant
	M30;	The end of the program

Task 3　Hands-on Training

3.7　Equipment and appliances

Equipment: CK6150e CNC lathe.
Cutters: the outer circle turning tool, the cut-off tool with a blade width of 3mm.
Fixture: the self-centering three-jaw chuck.
Tools: chuck wrenches, tool holder wrenches, etc.
Gauges: the 0~150mm vernier caliper and the 0~25mm outside micrometer.
Blank: $\phi 50 \times 95$.
Auxiliary appliances: cutter shim, brushes and so on.

3.8　Check before powering on

Refer to Table 3-5 based on spot-inspection to check the machine status.

Table 3-5　Preparing card for machine start-up

Check item		Test result	Abnormal description
Mechanical part	Spindle		
	Feed part		
	Tool holder		
	three-jaw chuck		
Electrical part	Main power supply		
	Cooling fan		
CNC system	Electrical components		
	Controlling part		
	Driving section		
Auxiliary part	Cooling system		
	Compressed air system		
	Lubricating system		

3.9　Preparation before machining

Before machining, the tools required for this task should be prepared and installed correctly. The origin of the workpiece is set according to the process requirements, and the CNC machining program is entered and graphic verification is performed.

3.10　Part machining

After verifying that there are no issues with the graphic verification process, the parts machining can be carried out. Before the parts are processed, you should understand the safety operation requirements of the machine tool in detail, and wear labor protection clothing and utensils. When processing parts, you should be familiar with the functions and positions of the operation buttons of the CNC lathe, and understand the methods of dealing with emergency situations. During the machining process, especially before cutting, the actual distance between the cutter and the workpiece should be observed, and the remaining movement amount displayed in the "Remaining Movement Amount" column on the display screen should be referenced. When the difference between the actual distance and the remaining movement amount is too large, the vehicle should be stopped and checked immediately to avoid collision. If there is any abnormality, the machine tool should be stopped in time.

3.11　Part inspection

After the parts are processed, the workpiece should be carefully cleaned, and in accordance with the relevant requirements of quality management, the processed parts should be subject to relevant inspections to ensure the production quality. Table 3-6 shows the commonly used "three-level" inspection cards for machined parts.

Table 3-6　"Three-level" inspection cards for machined parts

Part drawing number		Part name		Working step number	
Material		Inspection date		Working step name	
Inspection items	Self-inspection result	Mutual inspection result	Professional inspection		Remark
Conclusion	☐ Qualified　☐ Unqualified　☐ Repair　☐ Concession to receive Inspection signature: Date:				
Non-conforming item description					

Project Summary

As a typical machining part of CNC lathes, the drive shaft is widely used in various equipments. According to equipment conditions and precision requirements, there will be some differences in the processing technology. Programmers and operators need to formulate the processing technology reasonably according to the processing conditions to improve the processing accuracy of the parts and the production efficiency.

Exercises After Class

1. Fill in the blanks

(1) According to the application, screw threads can be divided into _____, _____, _____ and _____.

(2) The single canned cycle of inner and outer circle cutting of the CNC lathe of FANUC system is specified by the instruction _____, and the end face cutting cycle is specified by the instruction _____.

(3) If the pitch, P, is not marked, it is indicated as _____, and if it is _____, the pitch needs to be marked.

(4) The instruction G92 is mainly used for processing _____ and _____ with equal pitch.

(5) We often look up the table to determine the feed amount for threading, and the feed amount needs to follow the _____ principle.

2. True or false

(1) Thread cutting cycle instruction G92 can only be used to turn parallel threads, but not tapered threads. ( )

(2) During the execution of G92 instruction, the feed rate override knob and spindle speed override knob on the machine panel are invalid. ( )

(3) Instruction G92 is the modal code. ( )

(4) In the selection of thread measuring tools, the external thread is mainly detected by the thread ring gauge, and the internal thread is detected by the thread plug gauge. ( )

(5) In the instruction format of G92, F represents the feed rate. ( )

3. Choice questions

(1) For parts with many and complex surfaces, the process is often divided ( ).
    A. according to the cutter used   B. by number of installations
    C. by processing part   D. by roughing and finishing

(2) When determining the coordinate axes of the CNC machine tool, generally ( )

should be determined first.

    A. X-axis   B. Y-axis   C. Z-axis   D. A-axis

(3) In the following statements, (   ) is wrong.

    A. Each CNC program stored in the system must have a program number.

    B. A block is composed of one or more instructions, which represent all the actions of the CNC machine tool.

    C. In most systems, the block number is only used as the target position indication of "jump" or "program retrieval".

    D. The program comments of the FANUC system are enclosed in "()".

(4) Among the following instruction options, (   ) is not used for thread machining.

    A. G32   B. G90   C. G92   D. G71

(5) Use FANUC system instruction "G92 X(U) Z(W) F; " to process double-start thread, and then "F" in this instruction means (   ).

    A. thread lead   B. thread pitch   C. feed per minute   D. thread start angle

4. Short answer questions

(1) Briefly describe the types of threads.

(2) Briefly describe measurement methods for screw threads.

(3) Briefly describe the application scenario in which the thread cutting cycle instruction, G92, mainly used for.

5. Comprehensive programming questions

According to the requirements of the driven shaft drawing (Figure 3-5), CNC machining program of drive shaft should be written and the parts machining should be carried out.

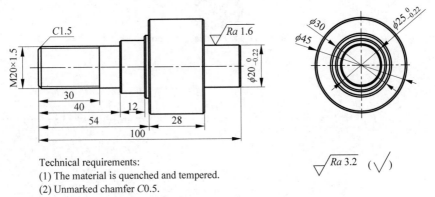

Technical requirements:
(1) The material is quenched and tempered.
(2) Unmarked chamfer C0.5.

Figure 3-5   the driven shaft

Self-learning test score sheet is shown in Table 3-7.

Table 3-7 Self-learning test score sheet

Tasks	Task requirements	Score	Scoring rules	Score	Remark
Learn key knowledge points	(1) Learn about the classification of threaded shafts and their machining characteristics (2) Thread detection gages can be used correctly (3) Master the use of threading cycle instructions and understand the meaning of each parameter	20	Understand and master		
Technological preparation	(1) Ability to read part drawings correctly (2) Ability to independently determine the process route and fill in the process documents correctly (3) Be able to write the correct processing program according to the processing technology	30	Understand and master		
Hands-on training	(1) Be able to reasonably select gages according to the structural characteristics and accuracy of the parts, and correctly standardize the measurement of relevant dimensions (2) Master the operation process of drive shaft turning process (3) Can correctly operate the CNC lathe and adjust the processing parameters according to the processing situation	50	(1) Understand and master (2) Operation process		

Ideological and Political Classroom

Project 4 Programming and Machining Training for Drive Wheel Milling

➢ Mind map

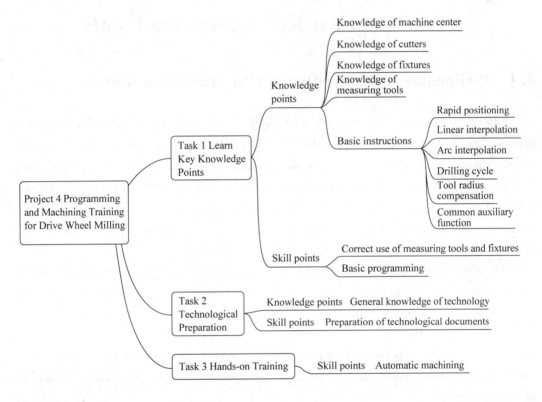

➢ Learning objectives

Knowledge objectives

(1) Be able to recognize the drawing of basic plane parts.

(2) Know the use and key requirements of the drive wheel in this mechanism.

Ability objectives

(1) Master the use of basic preparation instructions and auxiliary instructions.

(2) Master the selection and installation method of milling tools.

(3) Be able to independently determine the process routine and fill in the technological documents correctly.

(4) Be able to operate the CNC lathe correctly and adjust the machining parameters according to the machining conditions.

(5) Be able to select measuring tools reasonably according to the structural characteristics and accuracy of parts, and be able to measure the relevant dimensions correctly and normatively.

Literacy goals

(1) Cultivate students'scientific spirit and attitude;

(2) Cultivate students'engineering awareness;

(3) Develop students'teamwork skills.

Task 1　　Learn Key Knowledge Points

4.1　Preliminary understanding of CNC machining center

Computer Numerical Control milling machine, also known as CNC milling machine, means a milling machine controlled by computer numerical signals. It is an automatic processing equipment developed on the basis of general milling machine. The processing technology of the two is basically the same, and the structure is also somewhat similar. CNC milling machines are divided into two categories: without tool magazine and with tool magazine. Among them, the CNC milling machine with a tool magazine is also called a machining center.

After the workpiece is clamped once on the machining center, the digital control system can control the machine tool to automatically select and replace the tool according to different processes, automatically change the spindle speed, the feed rate, the movement path of the tool relative to the workpiece and other auxiliary functions, and complete the multi-process machining on several surfaces of the workpiece in turn. There are a variety of tool change or tool selection functions, so that the production efficiency is greatly improved.

Vertical machining center refers to the machining center whose spindle axis is perpendicular to the workbench. It is mainly suitable for processing complex parts such as plates, discs, molds and small shells. The vertical machining center can complete milling, boring, drilling, tapping and thread cutting. The vertical machining center is at least three-axis two linkage, and generally can realize three-axis three linkage. Some can perform five-axis and six-axis control. The column height of the vertical machining center is limited, and correspondingly the processing range of box type workpieces is reduced, which is the disadvantage of the vertical machining center. However, it is convenient for workpiece clamping and positioning. The movement track of the cutting tool is easy to observe, and the debugging program is convenient to check, so the problems can be found in time, and the machine can be stopped for processing or modification. The cooling condition is easy to reach, and the cutting fluid can directly reach the tool and the machining surface. The three coordinate axes coincide with the Cartesian coordinate system, so that intuitive feeling is the same as the drawing. The chips are easy to be removed and dropped to avoid scratching the processed surface. Compared with the corresponding horizontal machining center, it has simple structure, smaller floor area and lower price.

For the AVL650e vertical machining center (as shown in Fig. 4-1), the bottom of the column is designed with an "A" type high rigidity structure, the bed is made of high-strength high-grade cast iron structure, and the slide rails of each axis are fully supported. The three-axis slide rail adopts linear rolling guide and C3 grade precision ball screw, which has good positioning accuracy. The main motor directly drives the high-rigidity spindle with H.T.D toothed belt to make it move at high speed, which is up to 10000r/min. The tool arm type automatic tool changer has a capacity of 24 tools, and the tool changing time is only 2.5 seconds (tool to tool). Equipped with a spiral chip conveyor and a chip collector as standard, which is easy to remove chips and reduces the auxiliary time for removing iron chips. The electric box is equipped with a heat exchanger, which effectively reduces the failure rate of electrical components and improves the service life of the machine tool. In order to ensure machining accuracy, it is equipped with spindle oil temperature cooling device as standard.

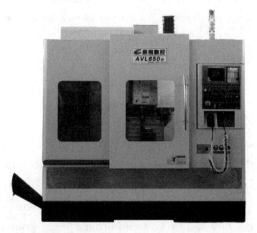

Figure 4-1　AVL650e vertical machining center

The main performance parameters of AVL650e are shown in Table 4-1.

Table 4-1　The main performance parameters of AVL650e

Item	Unit	Technical specifications
$X/Y/Z$ stroke	mm	620/520/520
Distance from the nose end of the spindle to the workbench	mm	100~620
Distance from spindle center to column slide rail surface	mm	540
Workbench size	mm×mm	800×500

Continued

Item	Unit	Technical specifications
Maximum load of the workbench	kg	500
T-groove size (width/pitch/number)	mm×mm×mm	18×130×3
Spindle speed	r/mim	100～10000
Spindle taper		BT40
Rapid feed rate ($X/Y/Z$)	m/min	48/48/48
Cutting feed rate ($X/Y/Z$)	mm/min	1～20000
Tool magazine capacity	pcs	24
Tool change method		Tool arm type
Tool change time (tool-to-tool)	s	2.5
Power capacity	kV·A	20
Machine size	mm×mm×mm	2400×2300×2700
Machine quality	kg	4200
Positioning accuracy	mm	0.01/full length
Repetitive positioning accuracy	mm	0.005
Air pressure requirements	bar①	6
CNC system		FANUC 0i-MF

4.2 Milling characteristics

Milling is a kind of cutting method which takes the rotary motion of the milling cutter as the main motion and takes the milling cutter or workpiece as the feed motion. It is characterized by:

(1) The multiple-cutting-edge tool is used for machining, and the cutting edges are alternately cut, so the tool has good cooling effect and high durability. However, compared with turning, the cutting process of milling is discontinuous, and the cutting layer parameters and cutting force are changed, which is easy to cause shock and vibration, thus affecting the improvement of processing quality.

(2) Milling has high production efficiency and wide processing range. Various milling cutters can be used on ordinary milling machines to complete processing planes (parallel plane, vertical plane, inclined plane), steps, grooves (right angle groove, V-shaped groove, T-shaped groove, dovetail groove and other special grooves) and special surfaces. In addition, the use of milling machine accessories such as indexing head can also complete the milling of spline shaft, screw shaft, tooth clutch and other workpieces.

(3) Milling has high machining accuracy. The economic machining accuracy is generally IT9～IT7, and the surface roughness Ra value is generally 12.5～1.6μm. The fine milling accuracy can reach IT5, and the surface roughness Ra value can reach 0.20μm.

---

① 1bar=100kPa.

Compared with turning, the definition of milling parameters is slightly different.

As shown in Fig. 4-2, the cutting speed in milling is usually expressed by the spindle speed. The spindle speed (n) is the number of revolutions per minute of the milling cutter on the spindle, and the unit is r/min. The feed rate (F) is usually expressed by the relative movement distance between the workpiece and the tool center per minute, and the unit is mm/min. The cutting depth can be divided into axial cutting depth (a_p) and radial cutting depth (a_e), which respectively represent the amount of metal removed by the tool on the workpiece surface and the width that the milling cutter diameter participates in the part cutting along the radial direction, and the unit is mm.

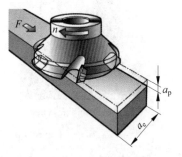

Figure 4-2 Milling models

4.3 Knowledge of cutters

There are many kinds of milling cutters. Generally, according to the processing characteristics, they can be divided into face milling cutter (plain milling cutter), shoulder milling cutter (side milling cutter), profiling milling cutter, groove milling cutter, thread milling cutter, chamfer milling cutter and gear milling cutter, etc., as shown in Figure 4-3.

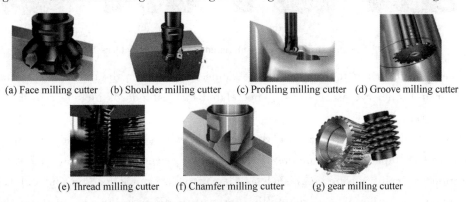

(a) Face milling cutter   (b) Shoulder milling cutter   (c) Profiling milling cutter   (d) Groove milling cutter

(e) Thread milling cutter   (f) Chamfer milling cutter   (g) gear milling cutter

Figure 4-3 Commonly used milling cutters

The structures of commonly used milling cutters include integral milling cutters, welding pad milling cutters and insert milling cutters, as shown in Figure 4-4. Integral milling cutters are made of the same material from the body to the end of the blade, most of which is high-speed steel, hard alloy or cermet, and most of them are milling cutters with small diameters or special shapes. The shank and body of the welding pad milling cutter are made of low-cost carbon tool steel, and the cutting edge is brazed with high-speed steel, carbide, CBN, diamond, etc. Insert milling cutters are these types of milling cutters in which the blade is fixed to the cutter body with screws or clips.

During the use of CNC milling tools, attention should be paid to the installation mode

(a) Integral milling cutter (b) Welding pad milling cutter (c) Insert milling cutter

Figure 4-4　The structures of commonly used milling cutters

of the tool holder on the spindle of the machine tool, that is, the type of the tool holder. Improper use of the tool holder may damage the tool holder or even the spindle of the machine tool. The commonly used tool holder types in daily work are shown in Figure 4-5.

(a)　　　(b)　　　(c)　　　(d)

Figure 4-5　The types of commonly used tool holders

The tool holder is a tool, which is the connection between the mechanical spindle and the cutter and other accessory tools. The related main standards include BT, JT (SK), CAPTO, BBT, HSK.

The BT tool holder realize the connection between the cutter and the spindle by adopting 7∶24 taper for the spindle of the machining center. The standard 7∶24 taper connection has many advantages, such as, it can realize quick loading and unloading of tools; under the action of the axial pulling force of the tie rod, the cone of the tool holder tightly contacts the inner conical surface of the spindle, so the solid cone directly supports the cutter in the conical hole of the spindle, which can reduce the overhang of the cutter; only one dimension needs to be machined to high accuracy, so the cost is low and reliable.

JT tool holder is equivalent to the evolution version of BT tool holder, which is called SK tool holder in Germany. JT tool holder has one more interface with the spindle than BT tool holder, but the positioning gap is unfavorable to the dynamic balance performance at high speed, and it needs to be improved by deduplication and other methods. The differences between the JT tool holder and the BT tool holder include the following points. The keyways for JT and BT tool holders, that meet the keys of the spindle end faces, are different. The keyway of the BT tool holder is half cut, and that of the JT tool holder is a through slot; JT tool holder has a notch on the grasping ring, but BT tool holder does not have it. The sizes of the two standard grasping rings are different, and the thickness of the tool changing gripper is different. Therefore, the two kinds of tool holders can not be interchanged during operation.

HSK tool holder is a tool holder that high-speed cutting technology is applied to cutting tools. High-speed cutting technology has become an important part of machining and manufacturing technology. HSK tool holder is a new type of high-speed short-tapered tool holder. Its interface adopts the method of simultaneous positioning of the tapered surface and the end surface. The tool holder is hollow, the length of the cone is short, and the taper is 1 : 10, which is conducive to realize light-weight and high-speed tool change. Due to the use of hollow cone and end face positioning, the radial deformation difference between the spindle hole and the tool holder during high-speed machining is compensated, and the axial positioning error is eliminated completely, making high-speed and high-precision machining possible. This tool holder is more and more common in high-speed machining centers.

4.4 Knowledge of fixture

Milling machine fixtures are mainly used for machining planes, grooves, splines and various forming surfaces on parts, and are one of the most commonly used fixtures. It is mainly composed of positioning device, clamping device, clamping body, connecting element and tool setting and guide element. When milling, the cutting force is higher, meanwhile, the vibration is higher due to intermittent cutting, so the clamping force of the milling machine fixture is required to be larger, and the stiffness and strength requirements of the fixture are relatively higher.

Like other machine tool fixture systems, milling machine fixtures are divided into general fixtures, special fixtures, adjustable fixtures, combined fixtures, etc. According to the clamping mode, they can also be divided into manual clamps, pneumatic clamps, hydraulic clamps, electric clamps, magnetic clamps and vacuum clamps, etc. Figure 4-6 shows the manual fixture commonly used in machining center.

(a) parallel vice　　　　　(b) chuck　　　　　(c) indexing head

Figure 4-6　Manual universal fixtures commonly used in machining center

The parallel vice is a general accessory for machine tools, which is used in conjunction with the workbench to fix, clamp and position the workpiece during processing. It is composed of body base, movable jaw, nut, screw rod and other components. According to its structure and use, it can be divided into general parallel vice, angle pressing parallel vice, tilting parallel vice, high precision parallel vice, force-enhanced parallel vice, etc.

A chuck is a mechanical device used to clamp a workpiece on a machine tool. It is a machine tool accessory that uses the radial movement of the movable jaws evenly distributed on the chuck body to clamp and position the workpiece. The chuck is generally composed of the chuck body, the movable jaw and the jaw drive mechanism. The diameter of the chuck body is a minimum of 65 mm and a maximum of 1500mm. There is a through hole in the center to allow the passage of workpieces or bars. The back has a cylindrical or short conical structure, which can directly connect with the machine tool through a flange, or can be pressed on the machine tool table through a pressure plate.

The indexing head is a machine tool accessory installed on the milling machine to divide the workpiece into any equal parts. The workpiece clamped between the centers or on the chuck can be divided into any angle by using the indexing scale ring, vernier, positioning pin, indexing plate and change gear, and the circumference can be divided into any equal parts. It assists the machine tool to process various grooves, spur gears, spiral spur gears, Archimedes spiral cams, etc. using various tools with different shapes. For the CNC machine tool, the CNC indexing head is driven by an AC or DC servo motor and actuated by a complex pitch worm gear set mechanism, It uses a hydraulic embracing locking device, coupled with a solid rigid sealing structure. CNC indexing heads are widely used in milling machines, drilling machines and machining centers. With the four-axis operation interface of the mother machine, simultaneous four-axis machining can be performed.

4.5 Knowledge of measuring tools

When installing fixtures or workpieces on milling machines, it is often necessary to ensure that a reference of fixtures or workpieces is parallel to the coordinate axis of the machine tool, i.e. "straightening". At this time, a dial indicator is required, as shown in Figure 4-7. Before some mold parts are processed, it is often necessary to use a dial indicator to align the workpiece coordinate system.

A lever arm test indicator, also known as a dial test indicator or finger indicator, is a measuring instrument that uses a lever-gear transmission mechanism or a lever-screw transmission mechanism to change the size into the angular displacement of the pointer and indicate the value of the length dimension. It is used to measure the geometric shape error and mutual position correctness of the workpiece, and the length can be measured by comparative method.

Figure 4-7 The dial indicators

4.6 Technology knowledge

The machining process is part of the mechanical production process and is a direct production process. It is the process of using mechanical processing methods to directly change the shape, size and surface quality of the blank to make it a qualified product part.

4.6.1 Positioning principle

When machining, the size, shape and positional accuracy of the workpiece are guaranteed by the relative position of the cutter and the workpiece. Before machining, the workpiece is installed at a certain position relative to the tool, which is called positioning.

Six-point positioning rule is expounded as follows. Any unconstrained object has six degrees of freedom in space. To make the object have a definite position in space, these six degrees of freedom must be constrained. The essence of workpiece positioning is to make the workpiece occupy a certain position in the fixture, so the problem of workpiece positioning can be transformed into the problem of determining the position of rigid body coordinates in the space Cartesian coordinate system. As shown in Figure 4-8, in the space Cartesian coordinate system, a rigid body has six degrees of freedom, that three degrees of freedom move along the X, Y, and Z axes and another three degrees of freedom rotate around these three axes. The six degrees of freedom of the workpiece are limited by six reasonably distributed support points, so that the workpiece occupies the correct position in the fixture, which is called the six-point positioning rule. When expounding the six-point positioning rule, people often use the example of milling non-through slots shown in Figure 4-8. Three points of a_1, a_2, and a_3 represent the main positioning surface A, which restricts the rotational degrees of freedom in the X and Y directions and the movement in the Z direction. The two points a_4 and a_5 represent the side B, which restricts the movement degree of freedom in the X direction and the rotation degree of freedom in the Z direction. Point a_6 embodies the thrust surface C and restricts the freedom of movement in the Y direction. In this way, all six degrees of freedom of the workpiece are restricted, which is called full positioning. Of course, positioning only ensures that the position of the workpiece in the fixture is determined, and does not guarantee that the workpiece does not move during processing, so it needs to be clamped. Positioning and clamping are two different concepts.

There are four positioning forms in production applications.

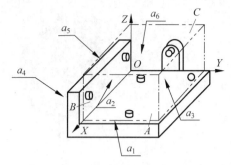

Figure 4-8　Six-point positioning principle

(1) Complete positioning: the positioning of the workpiece uses six support points, which limit all six degrees of freedom of the workpiece, so that the workpiece has a unique position in the middle of the fixture.

(2) Incomplete positioning: according to processing needs, only limits part of the degree of freedom of the workpiece.

(3) Under positioning: according to the requirements of processing technology, the degree of freedom that should be limited is not limited.

(4) Over positioning: positioning in which the same degree of freedom of the workpiece is repeatedly restricted by two or more different positioning elements.

4.6.2　Benchmarks

In the process of designing and manufacturing parts, to determine the positions of some points, lines or surfaces, some specified points, lines or surfaces must be used as the basis. These points, lines or surfaces are called benchmarks. According to the different functions, benchmarks are often divided into design benchmarks and process benchmarks.

Design benchmark is the benchmark of designing parts.

The process benchmark is the benchmark used in manufacturing parts, which is further divided into procedure benchmark, positioning benchmark, measurement benchmark, and assembly benchmark. The procedure benchmark is the benchmark used to demarcate the position of the machined surface in the process file. The positioning benchmark is a point, line or surface that is used to make the workpiece occupy the correct position on the machine tool or fixture. It is the most important benchmark in the process. Whether the selection of the positioning benchmark is reasonable plays a decisive role in ensuring the dimensional accuracy and shape and position accuracy of the workpiece after processing, arranging the processing sequence, improving the productivity and reducing the production cost. It is one of the main tasks in formulating the technological process. The positioning benchmark can be divided into rough benchmark and fine benchmark. The measurement benchmark is the benchmark used to measure the size and position of the machined surface. An assembly benchmark is a benchmark used to determine the location of a part or component in a machine.

The selection of the positioning benchmark is to ensure the positional accuracy of the workpiece. Therefore, its selection is always carried out from the surface that requires positional accuracy. The rough benchmark is the positioning benchmark of the rough surface, and generally the selection follows the following principles.

(1) Select the unmachined surface as the rough benchmark, so that the machined surface can have a more correct relative position, and it is possible to machine most of surfaces that to be machined in one installation.

(2) Select a surface that requires a uniform machining allowance as the rough benchmark, which can ensure that the surface as the rough benchmark is processed with a uniform margin.

(3) For surfaces where all surfaces to be machined, choose the surface with the smallest allowance and tolerance as the rough benchmark to avoid waste due to insufficient allowance.

(4) Select a smooth, flat and large surface as the rough benchmark.

(5) The rough benchmark should not be reused. In general, the rough benchmark is only allowed to be used once.

For parts that require high geometric tolerance accuracy, the machined surface should be used as the positioning benchmark, which is called a fine benchmark. Its selection generally follows the following principles.

(1) Benchmark coincidence principle: the principle of coincidence of positioning benchmark and design benchmark. As shown in Figure 4-9(a), select K as the fine benchmark to ensure dimension A when machining hole D.

(2) Principle of uniform benchmark: For the machined surfaces that require high positional accuracy, the same benchmark shall be used as much as possible. As shown in Figure 4-9(b), in order to ensure accuracy when machining gears, the inner hole $\phi 35$ and the end face B are uniformly used as the fine benchmark for each finishing process.

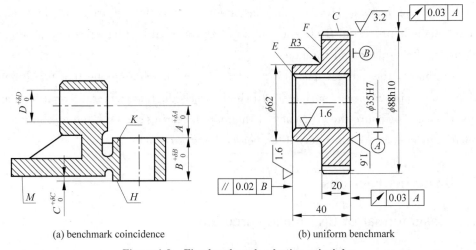

(a) benchmark coincidence   (b) uniform benchmark

Figure 4-9 Fine benchmark selection principles

(3) Mutual benchmark principle: among the surfaces to be processed, each other shall be used as the positioning reference during processing.

(4) Self-benchmark principle: take the machined surface itself as the positioning benchmark.

In a word, no matter the selection of rough or fine benchmark, the workpiece positioning must be stable, safe, and reliable first, and then the technical and economic principles such as easy fixture design, simple structure and low cost must be considered.

4.6.3 Principles for the formulation of machining process specification

The following principles shall be followed when formulating the machining process specification:

(1) Base surface first principle: when machining parts, a suitable surface must be selected as the positioning base surface so that the workpiece can be installed correctly. In the first process, only the blank surface (unmachined surface) can be used as the positioning base surface. In the subsequent process, in order to improve the processing quality, the machined surface should be used as the positioning base surface as much as possible. Obviously, when arranging the working procedure, the fine base surface should be machined first.

(2) The principle of separating rough and fine process and first rough process and then fine process. When the processing quality of parts is high, the surface with high precision requirements should be machined in different processing stages, which generally are three stages of roughing, semi-finishing and finishing. Finishing should be done last. In this way, it is beneficial to ensure the processing quality and the arrangement of some heat treatment procedures.

(3) The principle of first plane and then hole. For parts such as boxes and brackets, the plane should be machined first and then the hole should be machined. This is because the outline of the plane is flat, and the placement and positioning are stable and reliable. After the plane is processed first, the hole can be processed by plane positioning to ensure the position of the plane and the hole.

In summary, the general machining order is: first machining the fine benchmark → roughing the main surfaces (surfaces with high precision requirements) → finishing the main surfaces. To ensure the efficiency of the manufacturing system, the machining of the secondary surfaces is appropriately interspersed between stages.

4.7 Basic instructions

4.7.1 Basic format of CNC milling program

The basic format of a CNC milling program is shown in Table 4-2.

Table 4-2 Basic format of CNC milling program

Program details	Number	Instruction	Remark
Program header	①	O0001;	Program name
Program body	②	T01 M6;	Call the cutter
	③	S2000 M03;	Spindle forward rotation start
	④	G0 G90 G54 X__ Y__; G43 Z__ H1;	Move to (X__, Y__) in G54 coordinate system Move to the safe position in Z direction and start tool length compensation
	⑤	G0 X__ Y__ Z__;	Move to the starting point of the cutter
	⑥	Lowering the cutter, (feeding, cutting, retracting), raising the cutter	Milling the workpiece
	⑦	G0 Z__;	Return to Z direction safety position
	⑧	M5;	Spindle stop
	⑨		If other cutters are needed for processing, repeat the steps ②~⑧
Program end	⑩	M30;	End the program

The program head generally contains the program starter, the program name or program number. The program starter can be %, :, MP and other different identifiers according to different CNC systems. It can also be omitted in some CNC systems. Therefore, it needs to be specified according to programming manual of the machine tool when programming. In the FANUC 0i system, the program header can be directly represented by a program number with a capital letter "O" followed by a number. The program number is a positive integer with a value range of 0001~9999. The zero before the number can be omitted when inputting program number.

The program body is the main part of the program, which is the process of milling the workpiece with cutters. If there is only one cutter during the machining, the program body is ②~⑧. If there are multiple cutters, repeat the steps ②~⑧. Among them, the steps ②~④ are the initialization of the turning process. In these steps, the cutter is called, the feed rate is set, the workpiece coordinate system and the cutter compensation number are selected, and the cutter is moved to the safe position to prepare for the formal milling. The steps ⑤~⑥ is the formal milling process, which represents the machining process of an area on the workpiece. If an area needs to be milled multiple times, it only needs to repeat the process of "feeding, milling and retracting". If milling multiple areas, just repeat the steps ⑤~⑥. In the machining process of the vertical machining center, after each cutter has milled in all areas, at least the parts should be returned to the safe position of the Z direction, and the M5 command should be used to stop the spindle rotating.

The program end contains the end instruction and the end identifier of the program. The commonly used end instruction in the FANUC system is M02 and M30. Similar to the program header, the end identifier should be specified according to programming manual of the machine tool.

Commonly used preparatory function instructions and auxiliary function instructions are shown in Table 4-3.

Table 4-3 Commonly used preparatory function instructions and auxiliary function instructions

Instruction	Usage	Instruction	Usage
G00	Rapid positioning	G68	Coordinate system rotation (cancel with instruction G69)
G01	Line interpolation	G80	Drilling cycle canceled
G02	Clockwise circular interpolation	G8*	G81~G87 hole machining cycle
G03	Counterclockwise circular interpolation	M00	Program pause
G15	Cancel the polar coordinate mode	M03	Start Spindle-Clockwise
G16	Switch the polar coordinate mode	M04	Start Spindle-Counter Clockwise
G28	Return to reference point	M05	Stop Spindle
G40	Cancel tool radius compensation	M08	Turn on coolant
G41	Left compensation of tool radius	M09	Turn off coolant
G42	Right compensation of tool radius	M30	Program reset & rewind
G51	Scaling (cancel with instruction G50)	M98	Subprogram call
G52	Local coordinate system	M99	Subprogram return

4.7.2 Commonly used preparatory function instructions

1. Absolute coordinate instructions and relative coordinate instructions

There are two methods for specifying the movement of the cutter, which are absolute coordinate instructions and relative coordinate instructions. The absolute coordinate instruction is to specify the end point coordinate for the moving position of the cutter by the actual coordinate value of the workpiece coordinate system. The relative coordinate instruction is also called the incremental coordinate instruction, which is a method of specifying the end point coordinate in the form of increments relative to the current point of the cutter.

As shown in Figure 4-10, when moving the cutter from the start point to the end point, the program can be written in the following way.

(1) The program with the absolute coordinate instruction is:

G90 X40.0 Y70.0;

(2) The program with the relative coordinate instruction is:

G91 X-60.0 Y40.0;

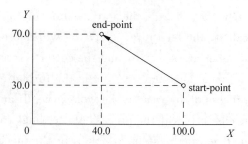

Figure 4-10 Absolute and relative coordinate instruction

2. Workpiece coordinate system

The workpiece coordinate system is the coordinate system used in programming, also known as the programming coordinate system, which is manually set for the convenience of programming. The FANUC system can specify 6 workpiece coordinate systems in G54~G59. The default coordinate system is G54 after startup. During machine operation, the workpiece origin offset value can be set through "Offset Setting" in the MDI unit. It can also be set through the program method or external data input method. As shown in Figure 4-11, in the application, it should be noted that the external workpiece origin offset value of the machine tool will affect the position of the origin of the G54~G59 workpiece coordinate system. In the actual operation of the machining center, the workpiece coordinate system can be set by the method of workpiece alignment in the machine.

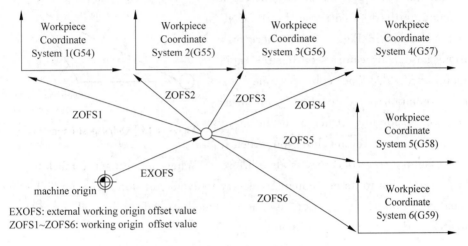

Figure 4-11 Workpiece coordinate system

3. Tool length compensation

As shown in Figure 4-12, by setting the offset between the virtual tool length during programming and the tool length used in actual machining in the offset memory, the offset of the tool length value can be compensated without changing the program. The FANUC system specifies the offset direction through the G43 command, and then uses the number (H code) after the address specified by the tool length compensation value to specify the tool length compensation value set in the offset memory.

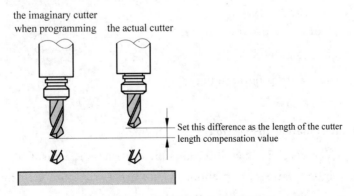

Figure 4-12 Tool length compensation

During operation, tool length compensation can be established by referring to the workpiece coordinate system or the machine tool coordinate system. This process is tool setting on the machine. It is also possible to determine the offset between the spindle reference point and the workpiece coordinate system when the workpiece is aligned, and the tool is calibrated by an external tool setter to import the actual length of the tool into the machine tool.

4. Quick positioning instruction

The instruction G00 enables the cutter to rapidly move from the current point to the next positioning point commanded by the block at the maximum feed speed preset by the CNC system. As shown in Figure 4-13, there may be various motion trajectories during the rapid positioning process. The common ones are linear interpolation positioning and non-linear interpolation positioning. Due to the different parameter settings of the machine tool when it

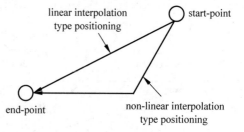

Figure 4-13 Quick positioning instruction

leaves the factory, the trajectory is also different. When operating the machine tool, you should pay attention to its movement trajectory to avoid collision accidents. The format of the rapid positioning instruction is the instruction code followed by the position coordinates of the positioning end point.

    G00 X__ Y__ Z__;

5. Linear interpolation instruction

The linear interpolation instruction, G01, is to move the cutter from its current position to the position required by the instruction in a straight line at the rate specified by the F code. The instruction format is:

    G01 X__ Y__ Z__ F__;

As shown in Figure 4-14, the tool cuts from point P_1 to point P_2 at a speed of 200mm/min. It is programmed with absolute coordinate instruction and relative coordinate instruction, respectively.

(1) Absolute coordinate programming:

    G01 G90 X20 Y18 Z-10 F200;

(2) Relative coordinate programming:

    G01 G91 X-20 Y-20 Z2 F200;

6. Circular interpolation instruction

G02 is the clockwise circular interpolation instruction, and G03 is the counter clockwise circular interpolation instruction. There are two common formats in the XY plane.

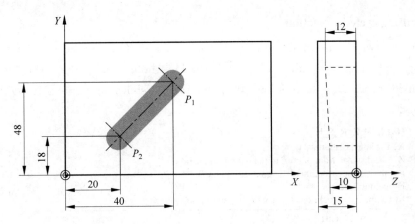

Figure 4-14 Linear interpolation instruction

(1) The method of specify radius method:

G02/G03 X __ Y __ R __ F __;

(2) The method of specify the center of the circle:

G02/G03 X __ Y __ I __ J __ F __;

The method of specifying the radius cannot directly process the whole circle. When the central angle of the processed arc is not greater than 180°, R takes a positive value, and when the center of the processed arc is greater than 180° and less than 360°, R takes a negative value.

The method of specifying the circle center can be used to process the whole circle and other arbitrary arcs. The values of *I* and *J* are the vectors from the starting point to the circle center, as shown in Figure 4-15.

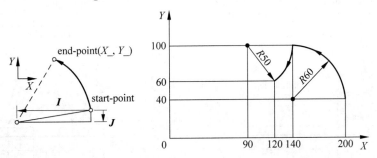

Figure 4-15 Circular interpolation instruction

(1) Programming with a specified radius:

G03 X140 Y100 R60;
G02 X120 Y60 R50;

(2) Programming with a specified center:

G03 X140 Y100 I-60 J0;
G02 X120 Y60 I-50 J0;

7. Drilling cycle instruction

There are two common drilling cycle instructions. One is the G81 instruction for processing general positioning holes and shallow holes, and the other is the G83 instruction for processing deep holes. Their formats are shown in Figure 4-16 and Figure 4-17 respectively.

G81 X_Y_Z_R_F_K_;
X_Y_: hole position data
Z_: the distance from point R to the bottom of the hole
R_: the distance from the datum plane to point R
F_: cutting feed rate
K_: number of repetitions (only when repetition is required)

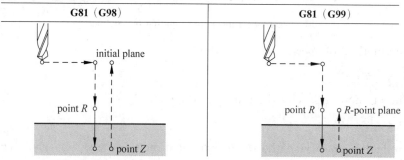

Figure 4-16　The drilling cycle instruction G81

G83 X_Y_Z_R_Q_,D_F_K_;
X_Y_: hole position data
Z_: the distance from point R to the bottom of the hole
R_: the distance from the datum plane to point R
Q_: feed amount each time
D_: retract amount
F_: cutting feed rate
K_: number of repetitions (only when repetition is required)

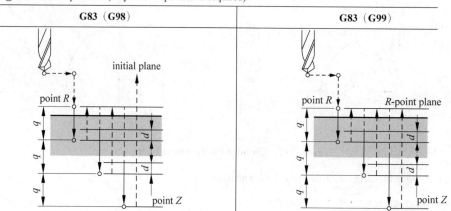

Figure 4-17　The drilling cycle instruction G83

The instruction G98 indicates that after machining a hole, the tool is lifted to the initial plane and then moved to the next hole. G99 indicates that after machining a hole, lift the tool to the plane specified by the R-point coordinate and then move to the next hole. In the actual operation of machining multiple holes, attention should be paid to there is no interference on the surface to avoid collision.

8. Tool radius compensation instruction

In the process of machining, the CNC machine tool controls the trajectory of the tool center. For convenience, the user always programming according to the part contour. Therefore, in order to process the required part contour, the tool center must be shifted to the inner side of the part by a tool radius value when machining the inner contour. During outer contour machining, the tool center must be offset to the outside of the part by a tool radius value. The function that the CNC device can automatically generate the tool center path in real time according to the program compiled according to the part outline and the preset offset parameters is called the tool radius compensation function. As shown in Figure 4-18(a), the solid line is the outline of the part to be machined, and the dashed line is the tool center path. According to the ISO standard, when the tool center path is on the right side of the programming path (part outline), it is called right tool compensation, which is realized by G42 instruction; otherwise, it is called left tool compensation and realized by G41 instruction.

As shown in Figure 4-18, in order to make the offset as large as the radius value of the tool, the CNC machine tool first creates an offset vector (starting tool) whose length is equal to the tool radius. The offset vector is perpendicular to the advancing direction of the tool. From the direction of the workpiece toward the tool center, if the linear interpolation or circular interpolation is specified after starting the tool, the tool can be processed after only offsetting a certain offset vector. Finally, in order to make the tool return to the starting point, cancel the tool radius compensation.

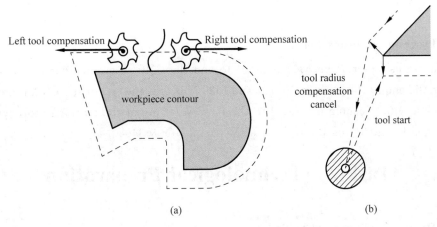

Figure 4-18 Tool radius compensation

In the XY plane, the programming format for tool radius compensation is:

G1 G41/G42 X__ Y__ D__

D is the tool compensation value code, which is usually the same as the tool number for easy identification. Use the G40 instruction to cancel tool radius compensation.

4.7.3 Commonly used auxiliary function instructions

1. Tool call

The vertical machining center uses M06 auxiliary command and T code to call the tool. The format is:

T＿ M06;

The value of T is the position code of the tool in the tool magazine.

2. Spindle function S code

The machining center uses S code to specify the rotation speed of the spindle, and the unit is r/min. When the value specified by S code is a positive integer and the value range is greater than the maximum rated speed of the machine tool, the spindle rotates at the maximum rated speed. The auxiliary instruction M code controls the start and stop of the spindle. M03 is the forward rotation of the spindle, M04 is the reverse rotation of the spindle, and M05 is the spindle stop. Usually, viewing from the negative direction of the Z-axis, the clockwise rotation of the spindle is defined as forward rotation. Its format is as following.

(1) Spindle forward rotation:

S＿ M03;

(2) Spindle reverse rotation:

S＿ M04;

(3) Spindle stop:

M05;

3. Program end instruction

You can use instruction M02 or M30 to end the CNC program. Among them, M02 indicates the end of the main program, and M30 indicates the end of the main program and reset. Its specific meaning is related to the factory setting of the machine tool, which can be redefined by modifying the parameter №3404#5 or №3404#4.

Task 2　Technological Preparation

4.8　Part drawing analysis

According to the use requirements of the part, 45 steel can be selected as the blank material, and the size is set as $\phi 255 \times 25$.

As shown in Figure 4-19, $\phi 47$ is the installation size of the spindle, and the precision can be obtained by fine boring on the lathe. During machining, the outer circle of $\phi 255$ and an end face are used as the rough benchmark, and the inner hole is finished. Because the $\phi 21$ hole is used for workpiece alignment in the subsequent processing, the accuracy is increased to IT8 (0.033mm). Then take $\phi 47$ and the finished end face as the benchmark, process $\phi 250$ and $\phi 160$, and ensure the thickness of 10mm in the two places.

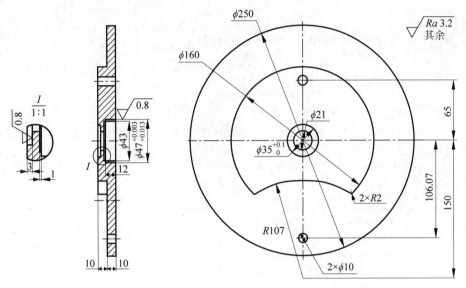

Figure 4-19　Part drawing of drive wheel

When milling, directly use $\phi 250$ as the benchmark. In order to ensure the accuracy, the inner hole of $\phi 21$ can be calibrated after alignment, and then the R107 and $\phi 35$ holes can be milled. Because the bottom of the $\phi 35$ hole has precision requirements, the sequence of roughing → semi-finishing → finishing should be adopted. $2 \times \phi 10$ are the mounting holes of the dial pin and the handle, and it adopts thread connection in the design drawing, so the accuracy requirement is not high, and it can be obtained directly by drilling.

Note that this part is a manual operation part, so after processed, the machining burr should be removed to ensure that the acute angle is sufficiently blunt to ensure personal safety during use.

4.9　Technological design

According to the analysis of the part drawing, the technological process is designed as shown in Table 4-4.

This training task is designed based on the 20th working procedure, namely milling, and the corresponding working procedure card is formulated as shown in Table 4-5.

Table 4-4 Technological process card

Machining process card	Product model	CLJG-01	Part number	QDL-01	Page 1	
	Product name	Geneva mechanism	Part name	Drive wheel	Total 1 page	
Material grade	C45	Blank size	$\phi 255 \times 25$	Blank quality	10kg	Quantity	1

| No. | Working Procedure | | | Work section | Technical equipment | Man-hour/min | |
	Name	Content				Preparation & conclusion	Single piece
5	Preparation	Prepare the material according to the size of $\phi 255 \times 25$		Outsourcing	Sawing machine		
10	Turning	Take the outer circle of $\phi 255$ and an end face as the rough benchmark, and finish the inner hole of $\phi 47$. The accuracy of $\phi 21$ hole is increased to IT8 (0.033mm)		Turning	Lathe, inner dial indicator	60	45
15	Turning	Positioning with $\phi 47$ inner hole, process the outer circle $\phi 250$ and the step $\phi 160$, guaranteed thickness of 10mm		Turning	Lathe, vernier caliper	90	30
20	Milling	Clamp with $\phi 250$ outer circle and calibrate the hole center of $\phi 21$. Mill the upper surface to ensure the total thickness of the part is 20mm. Mill $R107$ and $\phi 35$ and drill $2 \times \phi 10$ holes			Machining center, vernier caliper	90	90
25	Cleaning	Clean the workpiece, debur sharp corner		Locksmith			
30	Inspection	Check the workpiece dimensions		Examination			

Table 4-5 Working procedure card for milling

Machining working procedure card	Product model	CLJG-01	Part number	QDL-01	Page 1
	Product name	Geneva mechanism	Part name	Drive wheel	Total 1 page

Procedure No.	20
Procedure name	Milling
Material	C45
Equipment	Machining center
Equipment model	VAL6150e
Fixture	Three-jaw self centering chuck
Measuring tool	Vernier caliper
Preparation & conclusion time	90min
Single-piece time	90min

续表

Work steps	Content	Cutters	S/ (r/min)	F/ (mm/r)	a_p/ mm	a_e/ mm	Step hours/min	
							Mechanical	Auxiliary
1	Workpiece installation							5
2	Mill the upper surface of ϕ160, leave 0.1mm for thickness 20mm	ϕ63 face milling cutter	800	200	2	30	15	
3	Rough milling R107, leaving 0.1mm on the bottom and 0.2mm on the side	ϕ63 face milling cutter	800	200	2	30	20	
4	Fine milling ϕ160 surface	ϕ63 face milling cutter	1500	200	0.1	30	5	
5	Fine milling R107 bottom surface, leaving 0.2mm on the side	ϕ63 face milling cutter	1500	200	0.1	30	5	
6	Rough milling ϕ35, leaving 0.1mm on the side and bottom	ϕ12 end-milling cutter	2000	500	0.5	5	10	
7	Fine milling ϕ35 side	ϕ12 end-milling cutter	2000	200	0	0.1	5	
8	Fine milling ϕ35 bottom surface	ϕ12 end-milling cutter	2000	200	0.1	0	5	
9	Fine milling R107 side	ϕ12 end-milling cutter	2000	200	0	0.1	5	
10	Machining ϕ10 center hole	Center drill	1000	100	0.05	2	2	
11	Machining ϕ10 holes	ϕ10 Twist drill	1000	200	0.1	6	3	
12	R2 rounding							5
13	Disassemble and clean the workpiece							5

4.10 CNC machining program writing

According to the technology of the working procedure, the machining program is written as shown in Table 4-6.

Table 4-6 CNC machining program for drive wheel

No.	Program statement	Annotation
	O0001;	
	T01 M6;	Call ϕ63 face milling cutter
	G10 L13 P01 R0.2;	Set the side margin of 0.2mm
	S800 M3;	Set the spindle rotate forward at a speed of 800r/min
	G0 G90 G54 X115 Y0;	Locate to (X115, Y0)
	G43 Z10 H01;	Under the condition of length compensation number 01, cutter moves to position (X115, Y0, Z10)
N1	G1 Z0.1 F200;	Roughing the top surface

Continued

No.	Program statement	Annotation
	X80;	The first entry of the cutter
	G2 I-80 J0;	Cutting a cycle
	G1 X50;	The second entry of the cutter
	G2 I-50 J0;	Cutting a cycle
	G1 X20;	The third entry of the cutter
	G2 I-20 J0;	Cutting a cycle
	G1 X0;	The fourth entry of the cutter
	Z10;	
N2	G0 X0 Y-115;	Roughing $R107$
	G1 Z-2;	Process the first layer
	G1 G41 X79.738 Y-78.65 D1;	
	G3 X-79.738 R107;	
	G1 G40 X0 Y-115;	
	G1 Z-4;	Process the second layer
	G1 G41 X79.738 Y-78.65 D1;	
	G3 X-79.738 R107;	
	G1 G40 X0 Y-115;	
	G1 Z-6;	Process the third layer
	G1 G41 X79.738 Y-78.65 D1;	
	G3 X-79.738 R107;	
	G1 G40 X0 Y-115;	
	G1 Z-8;	Process the fourth layer
	G1 G41 X79.738 Y-78.65 D1;	
	G3 X-79.738 R107;	
	G1 G40 X0 Y-115;	
	G1 Z-9.9;	Process the fifth layer, leaving a margin of 0.1mm for the bottom surface
	G1 G41 X79.738 Y-78.65 D1;	
	G3 X-79.738 R107;	
	G1 G40 X0 Y-115;	
	Z10;	
N3	S1500 M3;	Finishing $\phi160$ surface
	G0 X115 Y0;	
	G1 Z0 F200;	
	X80;	
	G2 I-80 J0;	
	G1 X50;	
	G2 I-50 J0;	
	G1 X20;	
	G2 I-20 J0;	
	G1 X0;	
	Z10;	
N4	G10 L13 P01 R0.1;	Semi-finishing $R107$ side
	G0 X0 Y-115;	

Continued

No.	Program statement	Annotation
	G1 Z-10;	
	G1 G41 X79.738 Y-78.65 D1;	
	G3 X-79.738 R107;	
	G1 G40 X0 Y-115;	
	Z10;	
	M5;	
	T02 M6;	Call the second cutter, $\phi 12$ end-milling cutter
	G10 L13 P02 R0.2;	Setting radius compensates wear to 0.2mm, that is the side margin
	S2000 M3;	
	G0 G90 G54 X0 Y0;	
	G43 Z10 H2;	
N5	G1 Z-2 F500;	Roughing $\phi 35$
	G1 G41 X7.5 D2;	
	G3 I-7.5 J0;	
	G1 G40 X0 Y0;	
	G1 Z-2.9 F500;	
	G1 G41 X7.5 D2;	
	G3 I-7.5 J0;	
	G1 G40 X0 Y0;	
	G1 Z10;	
N7	G10 L13 P02 R0;	Finishing $\phi 35$ side
	G1 Z-2.9 F500;	
	G1 G41 X7.5 D2;	
	G3 I-7.5 J0;	
	G1 G40 X0 Y0;	
N8	G1 Z-3 F500;	Finishing $\phi 35$ bottom surface
	G1 G41 X7.5 D2;	
	G3 I-7.5 J0;	
	G1 G40 X0 Y0;	
N9	G0 X0 Y-115;	Finishing $R107$ side
	G1 Z-10;	
	G1 G41 X79.738 Y-78.65 D1;	
	G3 X-79.738 R107;	
	G1 G40 X0 Y-115;	
	G0 Z10;	
	M5;	
N10	T3 M6;	Machine the central hole
	S1200 M3;	
	G0 G90 G54 X0 Y65;	
	G43 Z10 H3;	
	G81 Z-1.5 R1 F100;	
	Y-106.07 Z-11.5 R-9;	

Continued

No.	Program statement	Annotation
	G80；	
	M5；	
N11	T4 M6；	Machine the central hole
	S1200 M3；	
	G0 G90 G54 X0 Y65；	
	G43 Z10 H4；	
	G83 Z-23 R1 Q5 F200；	
	Y-106.07 R-9；	
	G80；	
	M5；	
	G0 Z100；	
	M30；	

Task 3　Hands-on Training

4.11　Equipment and appliances

Equipment: AVL650e vertical machining center.
Cutters: $\phi63$ face milling cutter, $\phi12$ end-milling cutter, center drill, $\phi10$ twist drill.
Fixture: the K11-320 three-jaw self-centering chuck (with reverse claw).
Tools: kry file.
Measuring tools: 0~150mm vernier caliper and 0.02mm lever arm test indicator.
Blank: $\phi250\times23$ (after machining).
Auxiliary appliances: chuck wrench, rubber hammer, brush, etc.

4.12　Get to know the machine tool

4.12.1　Power on inspection

　　Check whether there is any abnormality in various parts of the appearance of the machine tool (such as the scatter shield, the footplate, etc.). Check whether the lubricating oil and coolant of the machine tool are sufficient. Check whether there are foreign objects on the tool holder, the fixture, and the baffle plate of the lead rail. Check the state of each knob on the machine tool panel is normal. Check whether there is an alarm after power on the machine tool. Refer to Table 4-7 to check the machine state.

Table 4-7 Preparing card for machine start-up

Check item		Test result	Abnormal description
Mechanical part	Spindle		
	Feed part		
	Tool holder		
	three-jaw chuck		
Electrical part	Main power supply		
	Cooling fan		
CNC system	Electrical components		
	Controlling part		
	Driving section		
Auxiliary part	Cooling system		
	Compressed air system		
	Lubricating system		

If the inspection of the machine tool is normal, the power of the machine tool can be turned on by rotating the electrical switch at the back of the machine tool.

Press the "power on" button on the control panel to power on the CNC system.

Rotate the emergency stop switch in the direction marked by the button to solve the emergency. The system panel after power on is shown in Figure 4-20.

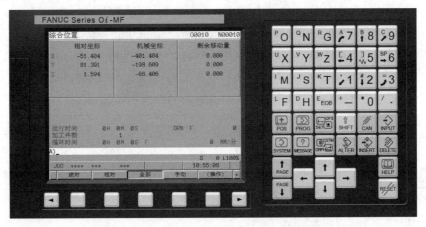

Figure 4-20 The system panel after powering on

4.12.2 Operation panel

Before formally operating the machine tool, you should be familiar with the functions and operation methods of each button on the operation panel of the machining center, and memorize the specific location of the emergency button.

1. Operation mode selection keys

The working mode selection keys are shown in Figure 4-21. After a certain working mode is selected, there will be a corresponding mark on the display screen.

(1) Connection mode (RMT): used for online processing or calling programs in CF card for processing.

(2) Edit mode (EDIT): used to edit programs or external CNC reading.

(3) Memory mode (MEM): used to automatically run programs read into memory.

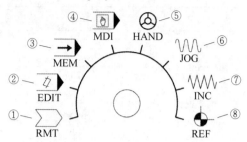

Figure 4-21 Operation mode selection keys

(4) Manual mode (MDI): used to run the program entered in the MDI panel of the controller.

(5) Handwheel mode (HAND): used to move the machine axes by the electronic handwheel.

(6) Fast moving mode (JOG): use the direction keys on the operation panel to quickly move the machine coordinate axis.

(7) Inching mode (INC): use the operation panel to jog the machine coordinate axis at a given distance.

(8) Return to origin (REF): used to return to the reference point of the machine tool.

2. Other function buttons

Other common function buttons on the operation panel are shown in Table 4-8. Due to different control systems, manufacturers, and factory batches, there will be some differences in the position and quantity of the panel function keys. Before operating the machine tool, you should be familiar with the operation manual of the current machine tool, and operate it in accordance with the requirements of safe operation.

Table 4-8 Other common operation functions on the panel

Icon	Description
MST Locking	Name: MST Locking button This function is only available in auto-correlation mode. (1) When the light of this button is on, the MST locking function is valid. After this function is turned on, when M, S, T auxiliary codes are programmed in the program, these codes are ignored and not executed, but the rest of the instructions can be executed normally. (2) When the built-in light of this button is off, the MST locking function is invalid. After this function is turned off, the M code, S code, and T code in the program are executed normally
Optional Skip	Name: Optional Skip button This function is only available in auto-correlation mode. (1) When the light of this button is on, the optional skip function is valid. When this function is turned on, in the case of auto running, if a "/" (slash) symbol is specified at the beginning of a block, this block is skipped and not executed. (2) When the light is off, the optional skip function is invalid. When this function is turned off, and there is a "/" (slash) symbol before a block, this block can be executed normally

Continued

Icon	Description
Single Block Execution	Name: Single Block Execution button This function is only valid in auto-correlation mode. (1) When the light of this button is on, the function of executing a single block is valid. When this function is enabled, the program will be executed in a single block. After the current block is executed, the program will be suspended. After continuing to press the "Program Start" button, the program of the next block can be executed, and so on. (2) When the light of this button is off, the function of executing a single block is invalid. The program will be executed until its end
Optional Stop	Name: Optional Stop button This function is only valid in the auto mode. (1) When the light of this button is on, the program optional stop function is valid. when this function is valid, if there is an M01 command in the execution program, the program will stop at this block. To continue executing the program, press the program start button. (2) When the light of this button is off, the program optional stop function is invalid. When this function is invalid, the program will not stop executing even if there is an M01 instruction in it
Z-axis locking	Name: Z-axis locking button (1) When the light of this button is on, the Z-axis locking function is valid. When this function is turned on, the Z-axis is locked and cannot be moved, but the relative coordinate/absolute coordinate of the Z-axis in the screen will change in real time with the command or manual movement. (2) After releasing this function, it is necessary to return to the mechanical zero point again. After this step is completed correctly, and then other related operations can be performed. (3) If the relevant operation is performed without returning to zero, it will cause coordinate offset, and even abnormal phenomena such as collision and program confusion, which will lead to danger. (4) When the light is off, the Z-axis locking function is invalid
Mechanical locking	Name: Mechanical locking button (1) When the light of this button is on, the mechanical locking function of all axes is valid. When this function is valid, no matter if any axis is moved in manual mode or automatic mode, the CNC will stop outputting pulses (movement commands) to the servo motor of this axis, but the command assignment is still performed, and the absolute and relative coordinates of the corresponding axis are also updated. (2) The M, S, T, B code will continue to be executed and is not restricted by mechanical locking. (3) After releasing this function, it is necessary to return to the mechanical zero point again. After this step is completed correcty, and then other related operations can be performed. (4) If the relevant operations are performed without returning to zero, it will cause coordinate offsets, and even abnormal phenomena such as collisions and program execution confusion, resulting in danger

Continued

Icon	Description
Empty Run	Name: Empty run button This function is only valid in auto-correlation mode (1) When the indicator light of this button is on, the empty run function is valid. When this function is turned on, the command to set the F value (cutting feed rate) in the program is invalid, and the movement rate of each axis is specified by the slow displacement rate (2) When the function is valid, if the program executes the cycle instruction, the feed rate cannot be changed by the slow feed rate or the cutting feed rate, and it still follows the F value in the control with a fixed feed rate displacement
Spindle Rotation-CW Spindle Stop Spindle Rotation-CCW	Name: Spindle rotation forward button, Spindle stop button, Spindle rotation reverse button (1) After executing the S code once on the machine, select the manual mode, and press the "Spindle rotation forward" or "Spindle rotation reverse" button, the spindle will rotate clockwise or counterclockwise. Spindle rotation speed $=$ the previously executed spindle speed S value $\times$ the gear where the spindle tune knob is located (2) Whether the spindle rotates forward or reverse, pressing the "Spindle stop" button can stop the rotating spindle (3) Conditions of use: ① They can only be used in "MDI" mode, "JOG" mode, and "INC" mode ② They are invalid in automatic operation
AUTO cooling Manual cooling	Name: Automatic cooling button In the automatic mode, When this button is pressed and the light is on, and the automatic operation of the coolant is valid. When the program executes to M08 or when this button is pressed, the coolant will spray out automatically, and it will automatically stop when the program executes to M09 Name: Manual cooling button In the MDI, JOG, and INC modes, when this button is pressed, the light and the coolant will be turned on. At this time, the light of the shift function button will flash
−X +X	Name: $+X/-X$ control button $+X/-X$ button: in the JOG mode, press and hold this button, the X-axis will move to the "$+/-$" direction (positive/negative direction) of the machine X-axis according to the speed of the feed override/rapid override, and the indicator lamp will turn on at the same time. When the button is released, the axis stops moving in the "$+/-$" direction of the machine X-axis, and the indicator lamp turns off at the same time In addition, when the X-axis direction movement instruction is executed in the program, the light of the button will also be on, and when the movement instruction is stopped, the built-in light will be off In addition, $+X$ button is also used as the X-axis zero return trigger button
−Y +Y	Name: $+Y/-Y$ control button $+Y/-Y$ button: in the JOG mode, press and hold this button, the Y-axis will move to the "$+/-$" direction (positive/negative direction) of the machine Y-axis according to the speed of the feed override/rapid override, and the indicator lamp will turn on at the same time. When the button is released, the axis stops moving in the "$+/-$" direction of the machine Y-axis, and the indicator lamp turns off at the same time In addition, when the Y-axis positive direction movement instruction is executed in the program, the light of the button will also be on, and when the movement instruction is stopped, the light will be off In addition, $+Y$ button is also used as the Y-axis zero return trigger button

Continued

Icon	Description
(−Z / +Z buttons)	Name: +Z/−Z control button +Z/−Z button: in the JOG mode, press and hold this button, the Z-axis will move to the "+/−" direction (positive/negative direction) of the machine Z-axis according to the speed of the feed override/rapid override, and the indicator lamp will turn on at the same time. When the button is released, the axis stops moving in the "+/−" direction of the machine Z-axis, and the indicator lamp turns off at the same time In addition, when the Z-axis positive direction movement instruction is executed in the program, the light of the button will also be on, and when the movement instruction is stopped, the light will be off In addition, +Z button is also used as the Y-axis zero return trigger button
O.T.REL.	Name: Overstroke release button (1) When the stroke of each axis of the machine tool exceeds the hardware limit, the machine tool will give an overstroke alarm. In the meanwhile, the light of this button will turn off, and the machine tool will stop. At this moment, press and hold this button and use the hand-held unit to move the overstroke axis in the opposite direction in handwheel mode (2) When the machine tool is powered on, release the emergency stop button, the machine tool starts normally without alarm, and the light of the button is always on
(Feed override knob)	Name: Feeding Override button and Feeding Override trimming button (1) This knob is located on the operation panel of this machine tool to control the programmed instruction G01 speed. Actual feeding speed = the value given by the F code in program × override value % where the feed override switch is located (2) In the INC mode, the function of this knob is to control the JOG feed override. The actual JOG feed speed = fixed value set by parameter × the override value % where the feed override switch is located (3) Used in conjunction with the axes feed control buttons
(Rapid override knob)	Name: Rapid Override button (1) This knob is located on the operation panel to control the programmed instruction G00 speed. Actual feeding speed = the maximum speed of parameter setting instruction G00 × rapid override value % where the rapid override knob is located (2) In JOG movement, its function is to control the manual rapid feeding override. The actual rapid feeding speed = the maximum speed of parameter setting instruction G00 × rapid the override value % where the rapid override knob is located (3) Used in conjunction with the axes feed control buttons
(Spindle trimming knob)	Name: Spindle trimming button (1) This knob is located on the operation panel to control the programmed spindle rotative speed, S. Actual rotative speed = the value given by the S code in program × the override value % where the spindle trimming knob is located (2) When the programmed speed exceeds the maximum rotative speed of the spindle, in other words, the override is more than 100%, the trimmed speed is equal to the maximum rotative speed of the spindle (3) Used in conjunction with the spindle control buttons

4.12.3 Return to origin

After the machine is started, each axis should be manually returned to the origin point. First, turn the operation mode knob to the "REF" mode, then press the "+Z" button on the operation panel to return the Z-axis to the origin point, and then press "+X" and "+Y" button respectively to return the X and Y axes to the origin point.

4.13 Cutters preparation

Before machining, the cutters required for this task should be prepared. Table 4-9 is a list of the cutters required. Install the cutters correctly according to the sequence in the Table 4-9.

Table 4-9 Cutters installation

Cutter No.	Tool holder	Rivet	Blade	Installation tool
T1				
T2				
T3				

Continued

Cutter No.	Tool holder	Rivet	Blade	Installation tool
T4				

When installing the tool, the correct installation tool and method should be used. Wrong operation may damage the cutter and the toolbar, and even cause personal injury. The installation accuracy of the cutter also has a great impact on the machining accuracy and cutter life.

4.14　Edge finder and workpiece alignment

4.14.1　Installing the edge finder

Use an ER chuck to install the mechanical edge finder on the BT tool handle, as shown in Figure 4-22. Then put the installed tool handle into the No. 24 cutter location of the machining center for standby.

The mechanical edge finder is a high-precision measuring tool that can quickly and easily set the precise center position of the mechanical spindle and the reference plane of the workpiece. The composition

Figure 4-22　Mechanical edge finder

of the mechanical edge finder mainly includes the following parts: hollow upper and lower detection heads of the clamping part and the eccentric part, springs that can penetrate the upper and lower detection heads, the top cover and the bottom cover which can hook and fix the spring respectively and are respectively embedded at the top and bottom ends of the upper and lower detection heads, etc. The working principle of the mechanical edge finder is that it is clamped on the machine tool and rotates at a low speed to align the machining center position automatically through eccentric adjustment.

4.14.2 The machining center calls the edge finder

First, switch the operation mode of the machining center to the MDI mode, and input the block "T24 M6;" in the MDI program input position of the system panel. Then press the cycle start key to run the current block, and the machine tool will execute the operation of calling the tool holder installed with the edge finder from the tool magazine.

Then input the block "S500 M3;", and press the cycle start key to run the current block, so that the edge finder rotates at a low speed of 500r/min.

4.14.3 Workpiece alignment

In the JOG mode, jog the X, Y, Z direction control keys to move the rotating edge finder to the left side of the workpiece. Switch to the HAND mode, and move the X-axis slowly with the handwheel to make the edge finder close to the workpiece until the upper and lower parts of the edge finder are concentric, as shown in Figure 4-23(a).

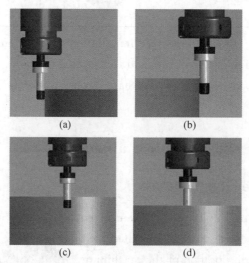

Figure 4-23 Alignment of the edge finder

Press the button ⎕ on the system panel, and select the button ⎕ at the bottom of the screen to display the relative coordinates of the machine tool. After inputting ⎕ ⎕, left click the button ⎕, and the X coordinate value in the relative coordinates becomes zero.

Only by moving the X and Z axes, make the edge finder reach the upper and lower concentric position on the right side of the workpiece, as shown in Figure 4-23(b). Record the coordinate value of X in the relative coordinates at this moment, here is 259.816, as shown in Figure 4-24. Divide it by 2 is 129.908.

Press the key [img] on the system panel, left click the key [img] at the bottom of the screen, enter "X129.908" in the X-axis position in the G54 coordinate system, and left click the key [img] to complete the workpiece alignment in the X direction.

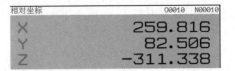

Figure 4-24   X-axis relative coordinate

In a similar way, move the Y direction of the edge finder to the side close to the operator, at this moment the relative coordinate value of X displays "X129.908". Slowly move the Y-axis direction with the handwheel to make the edge finder concentric up and down, as shown in Figure 4-23(c), and at this time preset the relative coordinate value of Y-axis to zero. Then, by moving only the Y and Z axes, the edge finder is in contact with the workpiece on the side away from the workpiece, and the upper and lower concentric states are achieved, as shown in Figure 4-23(d). Write down the displayed relative coordinate value of Y-axis at this moment, and input half of the value into the Y-axis position of the workpiece coordinate system. Then left click the key "Measure", the system will automatically calculate the offset value of the workpiece coordinate system G54.

Move upward in the Z-axis direction to make the edge finder completely away from the workpiece.

In order to verify the correctness of the workpiece alignment result, you can run the block "G0 G90 G54 X0 Y0;" in the MDI mode. At this time, the data displayed on the X and Y axes in the workpiece coordinate system (absolute coordinates) should be zero, and the tool center coincide with the workpiece center, as shown in Figure 4-25.

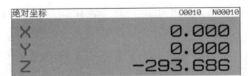

Figure 4-25   Verification after workpiece alignment

4.15   Cutter setting

Taking the first cutter as an example, the method of cutter setting in machining center is introduced.

Before setting the tool, the upper surface of the workpiece should be cleaned to ensure that there are no burrs and chips, and the zero position of the Z-axis setting instrument should be calibrated. The calibration height of the commonly used Z-axis setting instrument is 50mm. Place the Z-axis setting instrument on the workpiece at the position where the cutter is set, and ensure that the bottom surface is closely attached to the workpiece.

During tool setting, firstly, input the block "T1M6" in MDI mode to call out the first cutter, and use JOG mode to make it quickly approach the Z-axis setting instrument. When there is a certain distance from the top, the HAND mode is adopted, and the Z-axis is moved downward, so that the cutter teeth press down the movable block of Z-axis setting instrument. As shown in Figure 4-26, observe the value and stop moving downward when the pointer returns to zero. At this moment, the distance between the cutter teeth and the upper surface of the workpiece is 50mm.

Figure 4-26　Schematic of cutter setting

Press the button ▢ on the system panel of the machining center, and left click the button 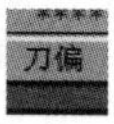 刀偏 at the bottom of the display. Move the cursor to the "shape (H)" position in the first row of the tool offset list, and input "Z10" in the data input box 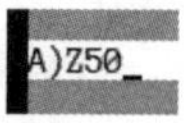 A)Z50_, and then left click the "Measure" button at the bottom of the display to complete the first cutter setting. The cutter length compensation setting of some machining centers does not have the "measurement" function. In this case, you only need to record the current value of Z-axis of the absolute coordinate, then subtract 50mm, and directly input the result to the corresponding position of "shape (H)".

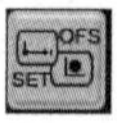

Note：

For milling cutters with multiple teeth and larger size, it should be subject to the tool setting data of the lowest tooth.

Where the length compensation data is input, the radius compensation value of the corresponding tool should also be output in the "Shape (D)" column for use in tool radius compensation.

In a similar way, the cutter setting operations of T2～T4 can be completed. The completed tool setting results are shown in Figure 4-27.

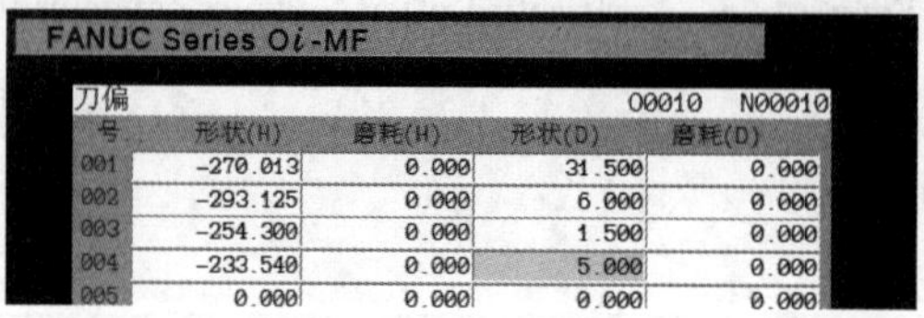

Figure 4-27　Cutter offset data for cutter setting

4.16　Program editing

There are two operations for program entry on the machine tool system panel, one is the entry of the program number, and the other is the entry of the program statement.

Switch the operation mode to the Edit mode on the operation panel, select the button [PROG] on the system panel, and then you can switch the program storage path through the "Directory" key at the bottom of the display screen [EDIT 程序 目录], and edit the program through the "Program" key.

When entering a program, first left click "Program" to enter the program editing interface.

First input the program number [P O][*0][*0][*0][;1], and directly press the key [INSERT], and then press the keys [E EOB] and [INSERT] to change the line. At this time, it should be noted that the program number (name) and the block end character are input in two steps. If you input "O0001;" and then press the key [INSERT], it will prompt "Format error" [A)O0001; 格式错误].

The input of other statements in the program is different from the program name. You can directly input the entire statement and then press the key [INSERT] directly. For example: [K T][;1][I M][SP →6][E EOB][INSERT].

The entered program is shown in Figure 4-28.

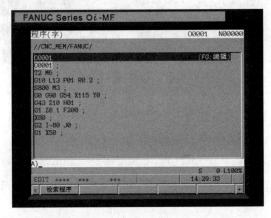

Figure 4-28 Entered CNC program

If a block needs to be deleted during the editing process, move the cursor to the position of the block to be deleted (highlighted in yellow), and then press the key [DELETE] to complete the deletion. If you need to modify the block selected by the cursor, you need to input the content to be changed at [A)_], and press the key [ALTER] to complete the replacement of the block content. To delete the last character in the input area [A)X100Y3000_], press the key [CAN] to backspace.

4.17 Simulation of CNC machining program

The FANUC-0i-MF system provides the graphic verification function for CNC machining programs. Due to the difference between the system version and the factory configuration of the machine tool manufacturer, the graphic verification interface also has its own differences.

Select "MEM" mode on the operation panel, also called the automatic execution mode of the program, and load the CNC machining program that needs graphic verification, and then press the key on the system panel to enter the graphic verification interface, where you can simulate the area size and viewing angle, as shown in Figure 4-29.

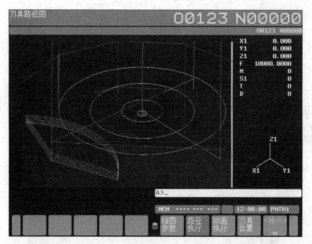

Figure 4-29 The graphic verification interface

Press the keys and on the operation panel, turn on the coordinate axis locking state and empty run state of the program, adjust the feed rate over ride key to 0%, press the program cycle start button, and the program will start graphic verification, and the verification speed can be adjusted by adjusting the feeding override.

4.18 Part machining

After the graphic verification process has verified that there is no problem, the parts machining can be carried out. Before the parts are processed, you should understand the safety operation requirements of the machine tool in detail, and wear labor protection clothing and utensils. When processing parts, you should be familiar with the functions and positions of the operation buttons of the CNC lathe, and understand the methods of dealing with emergency situations.

Note: If the graphic verification operation is performed, the machine tool must be returned to the mechanical zero point after the operation is completed, and then other related operations can be performed. If the relevant operations are performed without returning to zero point, it will cause coordinate offsets, and even abnormal phenomena such as collisions and program execution confusion, resulting in danger.

Select "MEM" mode on the operation panel, also called the automatic execution mode of the program, and load the CNC machining program that needs to be processed, and press the cycle start button to perform automatic processing.

The first automatic operation of the CNC machining program should be carried out in the debugging state. First turn down the machine feed rate override knob to 0%, and press the button 单节执行 on the operation panel, and then switch to the check interface in MEM mode by pressing the "Program Check" key at the bottom of the display, as shown in Figure 4-30.

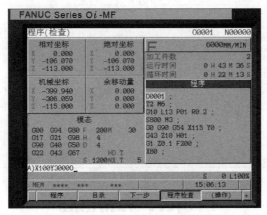

Figure 4-30　Check interface in MEM mode

In this state, the program will only automatically execute the line where the cursor is located each time when the cycle start button is pressed. Before the cycle start button is pressed, it should be observed whether the distance between the cutter and the workpiece is safe. After pressing the cycle start button, control the movement speed of the machine tool through the feed override knob, at the same time compare with the remaining movement amount displayed in the "Remaining Movement Amount" column of the display screen, therefore observe the actual distance between the cutter and the workpiece at the same time. If the difference between the actual distance and the remaining movement amount is too large, the operation should be stopped and checked to avoid a collision accident. In the process of program debugging, you should also pay close attention to the "modal" status on the display screen to ensure that there is no abnormality in the spindle revolution speed, feed speed, workpiece coordinate system number, compensation status and compensation number.

4.19 Part inspection

After the parts are processed, the workpiece should be carefully cleaned, and in accordance with the relevant requirements of quality management, the processed parts should be subject to relevant inspections to ensure the production quality. Table 4-10 shows the "three-level" inspection cards for machined parts.

Table 4-10 "Three-level" inspection cards for machined parts

Part drawing number		Part name		Working step number	
Material		Inspection date		Working step name	
Inspection items	Self-inspection result	Mutual inspection result	Professional inspection	Remark	
Conclusion	☐ Qualified ☐ Unqualified ☐ Repair ☐ Concession to receive Inspection signature: Date:				
Non-conforming item description					

Project Summary

Through the CNC milling of the driving wheel, master the basic format of CNC milling program and the use method of basic cutting instructions, and be able to use basic cutting instructions G1, G2, G3 to cut parts, and be able to use instructions G81 and G83 to process simple holes.

Master the basic operation methods of vertical machining centers, including: powering on and off, tool installation, workpiece alignment, cutter setting, program editing, graphic verification, CNC machining program debugging and automatic operation, etc.

Through task training, good professional quality, correct safety operation standards for machining centers, and basic machining quality awareness are developed and cultivated.

Exercises After Class

1. Fill in the blanks

(1) The biggest difference between the machining center and the general milling

machine is that the machining center can _____ .

(2) The machining procedures that can be completed by the machining center are _____ , _____ , _____ , _____ , etc.

(3) Compared with the corresponding horizontal machining center, the vertical machining center _____ , _____ and _____ .

(4) The automatic cutter changer of the machining center is composed of a drive mechanism and _____ .

(5) According to the use and processing method of the cutter, the forming turning tool belongs to the type of _____ .

2. True or false

(1) The height of the column of the vertical machining center is limited, and the processing range of the box type workpiece is smaller, which is the disadvantage of the vertical machining center. ( )

(2) The cutter change point is the place where the machining center manually changes the cutter. ( )

(3) In order to make the machine tool reach the thermal equilibrium state, the machine tool should run for more than 3 minutes. ( )

(4) In trial cutting and processing, after sharpening and changing the cutter, be sure to re-measure the cutter length and modify the cutter compensation value. ( )

(5) Using general computing tools and various mathematical methods to manually perform the operation of cutter path and perform instruction programming, which is called mechanical programming. ( )

3. Choice questions

(1) At present, the driving mode of the feeding system of the machining center is mostly adopted ( ).

  A. hydraulic servo feeding system   B. electric servo feeding system
  C. pneumatic servo feeding system   D. hydraulic-electrical combined type

(2) According to the types of spindles, machining centers can be divided into single axis machining center, double axes machining center, and ( ) machining center.

  A. non-replaceable headstock   B. three axes, five sides
  C. compound, four axes   D. three axes, replaceable headstock

(3) ( ) is a instruction for linear interpolation.

  A. G9   B. G111   C. G01   D. G93

(4) ( ) is a instruction for changing the cutter.

  A. G50   B. M06   C. G66   D. M62

(5) After the machine is powered on, you should first check if ( ) is normal.

  A. processing route
  B. each switch button and key

C. voltage, oil pressure and processing route

D. workpiece accuracy

4. Short answer questions

(1) Briefly describe the machining characteristics of CNC milling.

(2) Briefly describe fixtures used in the CNC machining center.

(3) Briefly describe the difference between the machining cycle instructions G81 and G83.

Self-learning test scoring sheet shows in Table 4-11.

Table 4-11 Self-learning Test Scoring Sheet

Tasks	Task requirements	Score	Scoring rules	Score	Remark
Learn key knowledge points	(1) Understand the structure and main parameters of AVL 650e vertical machining center (2) Understand the characteristics of milling processing (3) Familiar with the classification of commonly used milling cutters and can make the correct selection of tools (4) Understand the general fixtures and measuring tools of CNC milling machines (5) Master the basic format of CNC milling program and the use of basic preparation function instructions and auxiliary function instructions	25	Understand and master		
Technological preparation	(1) Correctly read basic planar part drawings (2) Be able to independently determine the processing technology and fill in the process documents correctly (3) Be able to write the correct processing program according to the processing process	25	Understand and master		

Continued

Tasks	Task requirements	Score	Scoring rules	Score	Remark
Hands-on training	(1) The corresponding equipment and utensils will be selected correctly (2) Master the selection and installation method of milling tools (3) Be able to reasonably select gages according to the structural characteristics and accuracy of parts, and correctly and normatively measure relevant dimensions (4) Can correctly operate the CNC machining center and adjust the processing parameters according to the processing situation	50	(1) Understand and master (2) Operation process		

Ideological and Political Classroom

Project 5 Programming and Machining Training for Geneva Wheel Milling

➢ Mind map

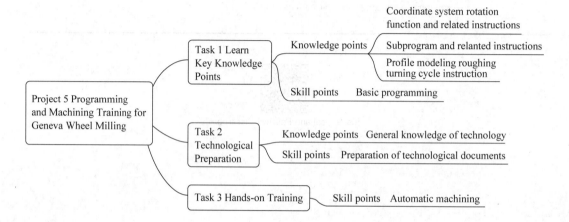

➢ Learning objectives

Knowledge objectives

(1) Know the working principle of Geneva wheel.

(2) Know the use and key requirements of Geneva wheel in this mechanism.

Ability objectives

(1) Master the use of coordinate system rotation instruction.

(2) Master the use of subprogram instruction.

(3) Master the selection and use of milling cutters.

(4) Be able to independently determine the process routine and fill in the technological documents correctly.

(5) Be able to operate the CNC lathe correctly and adjust the machining parameters according to the machining conditions.

(6) Be able to select measuring tools reasonably according to the structural characteristics and accuracy of parts, and be able to measure the relevant dimensions correctly and normatively.

Literacy goals

(1) Cultivate students' scientific spirit and attitude.

(2) Cultivate students' engineering awareness.
(3) Develop students' teamwork skills.

Task 1 Learn Key Knowledge Points

5.1 Function and principle of coordinate system rotation

In CNC programming, in order to describe the motion of the machine tool and to simplify the method of programming and ensure the interchangeability of recorded data, the coordinate system and movement direction of the CNC machine tool have been standardized. Both ISO and China have formulated the naming standards. The machine coordinate system is a rectangular coordinate system composed of X, Y and Z axes, which is established by taking the machine origin, O, as its origin and following the right-hand Cartesian rectangular coordinate system. The machine tool coordinate system is the basic coordinate system used to determine the workpiece coordinate system. It is an inherent coordinate system on the machine tool and has a fixed coordinate origin.

In practical application, it is often encountered that the features of some parts with the same geometric size are distributed around a certain point, and the engineering drawings of some parts use the rotating partial view to represent the geometric size. At this time, for the convenience of programming, the method of rotating the original workpiece coordinate system is often used for CNC programming.

For vertical machining centers, the commonly used rotation transformation is the transformation in the XOY plane. As shown in Figure 5-1, point A (x, y) rotating θ angle around point $O'(x_0, y_0)$ to $A'(x', y')$.

Assuming that the center of rotation is located at the origin of the coordinate system, after the coordinates (x, y) are rotated to (x', y'), the rotated coordinates can be easily obtained as

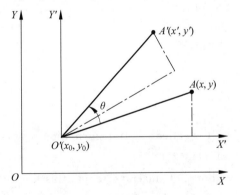

Figure 5-1 Machine coordinate system rotation

$$\begin{cases} x' = x\cos\theta - y\sin\theta \\ y' = x\sin\theta + y\cos\theta \end{cases}$$

This is a basic rotation transformation, written in matrix form as

$$[x' \quad y'] = [x \quad y] \begin{bmatrix} \cos\theta & \sin\theta \\ -\sin\theta & \cos\theta \end{bmatrix}$$

In practical application, the rotation center is often not the origin of the working coordinate system. As shown in Figure 5-1, the rotation center is at the position of point

$O'(x_0, y_0)$. Therefore, coordinate transformation is required for coordinate system rotation, that is, the origin of coordinate system is temporarily moved to the rotation center $O'(x_0, y_0)$. And then rotated according to the method when the rotation center is at the coordinate origin. When the coordinates are sent to the CNC system, the coordinate coefficient value is converted into the coordinate value in the original workpiece coordinate system.

After moving the coordinate center to the temporary position $O'(x_0, y_0)$, the coordinate value of the original point A can be expressed as

$$\begin{cases} x' = x - x_0 \\ y' = y - y_0 \end{cases}$$

This transformation is written in matrix form as

$$[x' \quad y' \quad 1] = [x \quad y \quad 1] \begin{bmatrix} 1 & 0 & 0 \\ 0 & 1 & 0 \\ -x_0 & -y_0 & 1 \end{bmatrix}$$

Therefore, in the XOY workpiece coordinate system, the coordinates of the rotated point $A'(x', y')$ are expressed geometrically as

$$\begin{cases} x' = (x - x_0)\cos\theta - (y - y_0)\sin\theta + x_0 \\ y' = (x - x_0)\sin\theta + (y - y_0)\cos\theta + y_0 \end{cases}$$

It can be expressed in matrix form as

$$[x' \quad y' \quad 1] = [x \quad y \quad 1] \begin{bmatrix} 1 & 0 & 0 \\ 0 & 1 & 0 \\ -x_0 & -y_0 & 1 \end{bmatrix} \begin{bmatrix} \cos\theta & \sin\theta & 0 \\ -\sin\theta & \cos\theta & 0 \\ 0 & 0 & 1 \end{bmatrix} \begin{bmatrix} 1 & 0 & 0 \\ 0 & 1 & 0 \\ x_0 & y_0 & 1 \end{bmatrix}$$

Then

$$\boldsymbol{T} = \begin{bmatrix} 1 & 0 & 0 \\ 0 & 1 & 0 \\ -x_0 & -y_0 & 1 \end{bmatrix} \begin{bmatrix} \cos\theta & \sin\theta & 0 \\ -\sin\theta & \cos\theta & 0 \\ 0 & 0 & 1 \end{bmatrix} \begin{bmatrix} 1 & 0 & 0 \\ 0 & 1 & 0 \\ x_0 & y_0 & 1 \end{bmatrix}$$

It is called the transformation matrix that rotates around the point $O'(x_0, y_0)$ in the XOY plane.

For CNC machine tools, this kind of transformation calculations are usually written into the mode calling macro program, and they are defined as the G codes of the preparation function. The corresponding G codes in FANUC $0i$ system is G68 and G69 instructions. The instruction format is:

Apply the coordinate rotation function and the format:

G68 X<x0> Y<y0> R<θ>;

Where (x_0, y_0) is the coordinate of the rotation center, θ is the rotation angle, and the counterclockwise direction is the positive direction.

Cancel the coordinate rotation function and the format:

G69;

As shown in Figure 5-2, when machining the right angle △ABC, according to the traditional programming method, it is necessary to use trigonometric functions to calculate the coordinate values of points B and C, and then perform cutting. When using the method of rotating coordinate system for programming, we can regard △ABC as its congruent △AB'C' rotated 30° counterclockwise around point A, and the coordinate values of points B' and C' can be easily calculated. The corresponding program is shown in Table 5-1.

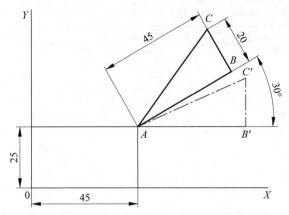

Figure 5-2 A programming example of coordinate rotation

Table 5-1 Programming example of coordinate system rotation instruction

Statement	Explanation
G0 X45 Y25;	Quickly locate to the starting point A
G1 Z-2 F500;	Cutting into the workpiece in the Z direction
G68 X45 Y25 R20;	Define the start of coordinate system rotation mode. Rotate 20° around point (45,25).
G1 X90;	Cut to point B'
Y45;	Cut to point C'
G69;	Cancel the coordinate system rotation mode
G1 X45 Y25;	Cut to point A

5.2 Format and application of subprogram

In a processing program, if some of the processing contents are identical or similar, in order to simplify the program, these repeated blocks can be listed separately and written into a separate program in a certain format, and named separately. This group of blocks is called a subprogram. If the main program needs a certain subprogram during the execution process, the subprogram can be called through the calling instruction. After the subprogram is executed, it will return to the main program and continue to execute the following blocks. Usually, a subprogram cannot be used as an independent machining program, it can only be called to realize local actions in machining.

5.2.1 Format of subprogram

There is no essential difference between the subprogram and the main program in the format, the main difference is the way of ending. The main program uses instruction M02 or instruction M30 to end the program, while the subprogram uses instruction M99 to end it and automatically return to the program that called it. As shown in Table 5-2, it is a comparison between a subprogram and a main program that has called the subprogram once.

Table 5-2 Format comparison between main program and subprogram

Main program	Subprogram
O0001;	O1234;
T1 M6;	G1 Z F2 F500;
S2000 M3;	G1 X80 F1000;
G0 G90 G54 X20 Y20;	AND80;
G43 Z10 H1;	X20;
M98 P0011234;	AND20;
G0 Z100;	G1 Z10;
M5;	M99;
M30;	

When naming a subprogram, the characteristic function specified by the system or the subprogram number individually defined by the machine tool manufacturer should be avoided.

5.2.2 Call subprogram

The common calling format of FANUC 0i system subprogram is:

M98 P○○○△△△△;

Where ○○○ is the number of calls, and the value range is 001～999. When inputting, the zero in front of the number can be omitted. If the number of calls is 1, "1" can also be omitted. △△△△ is the subprogram number. The value range is 0001～9999, and the zero before the number cannot be omitted when inputting.

When the number of calls is more, use the parameter L to specify the number of calls. The subprogram calling format is:

M98 P△△△△ L○○○○○○○;

When the subprogram number is also greater than 5 digits, the subprogram calling format is:

M98 P△△△△△△△ L○○○○○○○;

When the subprogram is not named with numbers but with characters, the subprogram calling format is:

M98 <△△△△> L○○○○○○○;

Subprograms can be called not only by the main program, but also by other subprograms. This method is called subprogram nesting, as shown in Figure 5-3. According to different systems, the nesting levels of its subprograms are also different. Generally, FANUC 0i system can be nested 4 levels, and newer systems such as FANUC 0i MF can be nested 15 levels.

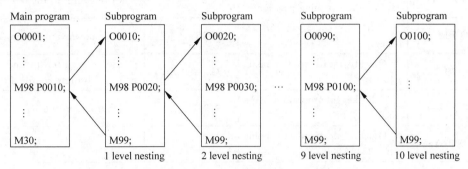

Figure 5-3 Schematic diagram of subprogram nesting of FANUC 0i MF system

5.2.3 Application of subprogram when a similar shape is processed repeatedly

When similar shapes are processed repeatedly, it is often involved that such positions starting from different starting points. Therefore, the positioning using the absolute coordinate mode is usually carried out in the main program. Shape repetition is realized in subprogram because relative coordinate programming is adopted. For processing the part shown in Figure 5-4, the subprogram programming can be used by referring to the method in Table 5-3.

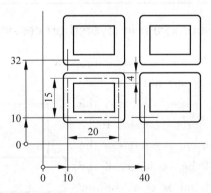

Figure 5-4 Programming of similar shape repeative machining

Table 5-3 Programming example of similar shape repeative machining

Main program		Subprogram	
O0001;	Main program number	O1234;	Subprogram number
T1 M6;	Call the cutter T1	G1 Z-2 F500;	Cutting into the workpiece in the Z direction
S2000 M3;	Spindle rotates forward at 2000r/min	G1 G91 X20 F1000;	Move 20mm to the right in the relative coordinate direction

Continued

Main program		Subprogram	
G0 G90 G54 X10 Y10;	Position to coordinate G54 absolute coordinate (10, 10)	Y15;	Move upward 15mm
G43 Z10 H1;	Length compensation number H01, to the starting position	X-20;	Move 20mm to the left
M98 P1234;	Call the subprogram O1234 once	Y-15;	Move downward 15mm
G0 X40 Y10;	Quickly locate to the point (40, 10)	G1 G90 Z10;	Lift the cutter to the start point in absolute coordinates
M98 P1234;	Call the subprogram O1234 once	M99;	Subprogram returns
G0 X40 Y32;	Quickly locate to the point (40, 32)		
M98 P1234;	Call the subprogram O1234 once		
G0 X10 Y32;	Quickly locate to the point (10, 32)		
M98 P1234;	Call the subprogram O1234 once		
G0 Z100;	Quickly lift the cutter to a safe position		
M5;	Spindle stops		
M30;	The main program ends and resets		

5.2.4 Application of subprogram in layered processing

As shown in Figure 5-5, during machining, the cutter cannot cut to 30mm at Z direction at one time. When programming is done with the cutting depth of each layer of 2mm, the XY surface profile of the same part needs to be repeated 15 times. If the contour is complex, the number of program lines will be too large. If the subprogram function is used, all parts of contour machining can be placed in the subprogram. In the first line of the subprogram, the cutting depth of each layer is specified in relative coordinates, and then the contour machining program is written in absolute coordinates. Do not write a cutter lift instruction at the end of the subprogram, and the coordinates (x, y) of the end point at the end of the subprogram should be the same as at the beginning of the subprogram.

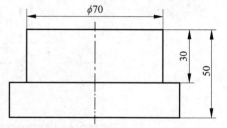

Figure 5-5 Layered machining programming

In the main program, the M98 instruction is used to call the subprogram and specify the number of calls. Note that the number of calls must be an integer, so it is necessary to flexibly control the cutting depth of each layer to ensure that the selected cutting depth of each layer can be divided by the total cutting depth.

int [number of calls] = total cutting depth ÷ cutting depth of each layer

Table 5-4 is an example of layered processing programming written in subprogram method.

Table 5-4 Example of layered machining programming

Main program		Subprogram	
O0001;		O1234;	
T1 M6;		G1 G91 Z-2 F500;	Each layer is cut off 2mm relative to the current position
S2000 M3;		G1 G90 G41 X45 Y10 D1;	Absolute coordinate mode, introducing tool compensation
G0 G90 G54 X60 Y0;	Select the outer point (60, 0) of the blank for cutting	G3 X35 Y0 R10;	Plunging into the contour with an R 10 arc
G43 Z10 H1;		G2 I-35 J0;	Cutting profile
G1 Z0 F500;	Move to the starting point of layered processing in advance	G3 X45 Y-10 R10;	Cut out the contour with an R 10 arc
M98 P151234;	Call the subprogram O1234 15 times	G1 G40 X60 Y0;	Cancel cutter compensation and return to the starting point.
G1 Z10;	Lift the cutter	M99;	Subprogram returns
G0 Z100;			
M5;			
M30;			

Task 2 Technological Preparation

5.3 Part drawing analysis

As shown in Figure 5-6, according to the use requirements of the part, 45 steel is selected as the blank material, and the blank size is $\phi 215 \times 25$.

$\phi 35$ is the key dimension of the part, which needs to be matched with the bearing, which can be obtained by boring during turning. The outer circle size of the parts is a large clearance fit in the whole machine tool, so rough turning can be done. $R80$ is a clearance fit around, which is obtained by rough milling. The 20-slot requires high positional accuracy and sliding friction with the dial pin, so it requires a low side surface roughness and requires fine milling.

Note that this part is a manual operation part, so after the process is finished, the machining burr should be removed to ensure that the acute angle is sufficiently blunt to ensure personal safety during use.

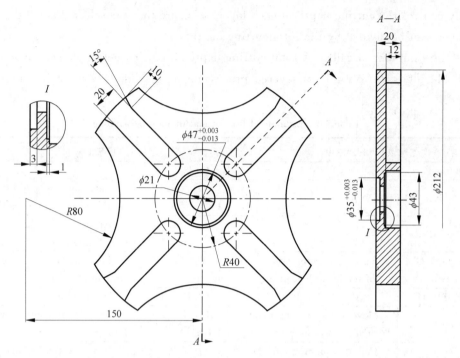

Figure 5-6 Parts drawing of Geneva wheel

5.4 Technological design

According to the analysis of parts drawing, the technologicalprocess is determined shown in Table 5-5.

Table 5-5 Technological process card

Machining process card	Product model			Part number		Page	
	Product name			Part name		Total page	
Material grade		Blank size		Blank quality	kg	Quantity	
Working step				Work section	Technical equipment	Man-hour/min	
No.	Name		Content			Preparation & conclusion	Single piece

The main work in this task is milling, which shall be designed in detail for the technological design. In combination with the training task 4, the technological design for the Geneva wheel milling shall be carried out, and the working procedure card shall be prepared in detail, as shown in Table 5-6.

Table 5-6 Working procedure card for milling

Machining working procedure card	Product model		Part number		Page	
	Product name		Part name		Total page	

							Procedure No.		
							Procedure name		
							Material		
							Equipment		
							Equipment model		
							Fixture		
							Measuring tool		
							Preparation & conclusion time		
							Single-piece time		

Work steps	Content	Cutters	S/ (r/min)	F/ (mm/r)	a_p/ mm	a_e/ mm	Stepping hours/min	
							Mechanical	Auxiliary

5.5 CNC machining program writing

According to the machining technology of the working procedure, the machining program is written as shown in Table 5-7.

Table 5-7 CNC machining program for Geneva wheel

Block No.	Program statement	Annotation
	O0001;	Main program
	T1 M6;	
	S1500 M3;	

Continued

Block No.	Program statement	Annotation
	G0 G90 G54 X120 Y0;	
	G43 Z10 H1;	
N1	G1 Z0 F500;	Position to milling starting point of $R80$ arc
	M98 P111001;	Call the subprogram to process the $R80$ arc to the depth Z-22. The subprogram is called 11 times
	G1 Z10;	
	G68 X0 Y0 R90;	Coordinate rotation 90° to machine the second one
	G1 Z0 F500;	
	M98 P111001;	
	G1 Z10;	
	G68 X0 Y0 R180;	Coordinate rotation 180° to machine the third one
	G1 Z0 F500;	
	M98 P111001;	
	G1 Z10;	
	G68 X0 Y0 R270;	Coordinate rotation 270° to machine the fourth one
	G1 Z0 F500;	
	M98 P111001;	
	G1 Z10;	
	G69;	
N2	G68 X0 Y0 R45;	Coordinate rotation 45°, roughing groove
	G0 X120 Y0;	
	G1 Z0 F500;	
	M98 P111002;	Call the groove machining subprogram 11 times, cut 2mm each time
	G1 Z10;	
	G68 X0 Y0 R135;	
	G0 X120 Y0;	
	G1 Z0 F500;	
	M98 P111002;	
	G1 Z10;	
	G68 X0 Y0 R225;	
	G0 X120 Y0;	
	G1 Z0 F500;	
	M98 P111002;	
	G1 Z10;	
	G68 X0 Y0 R315;	
	G0 X120 Y0;	
	G1 Z0 F500;	
	M98 P111002;	
	G1 Z10;	
	G69;	
N3	G68 X0 Y0 R45;	
	G0 X120 Y0;	

Continued

Block No.	Program statement	Annotation
	M98 P1003;	Finishing 45° groove
	G68 X0 Y0 R135;	
	G0 X120 Y0;	
	M98 P1003;	Finishing 135° groove
	G68 X0 Y0 R225;	
	G0 X120 Y0;	
	M98 P1003;	Finishing 225° groove
	G68 X0 Y0 R315;	
	G0 X120 Y0;	
	M98 P1003;	Finishing 315° groove
	G0 Z100;	
	M5;	
	M30;	
	%	
	O1001;	The subprogram for roughing $R80$ arc
	G1 G91 Z-2 F500;	
	G1 G90 G41 X100 Y62.45 D1;	
	G3 Y-62.450 R80;	
	G1 G40 X120 Y0;	
	M99;	
	%	
	O1002;	The subprogram for roughing groove
	G1 G91 Z-2 F500;	
	G1 X40;	
	X120;	
	M99;	
	%	
	O1003;	The subprogram for finishing groove
	G1 Z-21 F500;	
	G1 G41 X120 Y16.557 D1;	
	X95.527 Y10;	
	X40;	
	G3 Y-10 R10;	
	G1 X95.527;	
	X120 Y-16.557;	
	G1 G40 X120 Y0;	
	G1 Z10;	
	M99;	

Task 3　Hands-on Training

5.6　Equipment and Appliances

Equipment: AVL650e vertical machining center.

Cutters: $\phi16$ end-milling cutter.

Fixture: the K11-320 three-jaw self-centering chuck (with mandrel).

Tools: kry file.

Measuring tools: 0～150mm vernier caliper and 0.02mm lever arm test indicator (with mounting rod).

Blank: $\phi212\times20$ (after turning).

Auxiliary appliances: Chuck wrench, rubber hammer, brush, etc.

5.7　Check before powering on

Check whether there is any abnormality in various parts of the appearance of the machine tool (such as the scatter shield, the footplate, etc.). Check whether the lubricating oil and coolant of the machine tool are sufficient. Check whether there are foreign objects on the tool holder, the fixture, and the baffle plate of the lead rail. Check the state of each knob on the machine tool panel is normal. Check whether there is an alarm after power on the machine tool. Refer to Table 5-8 to check the machine state.

Table 5-8　Preparing card for machine start-up

	Check item	Test result	Abnormal description
Mechanical part	Spindle		
	Feed part		
	Tool holder		
	three-jaw chuck		
Electrical part	Main power supply		
	Cooling fan		
CNC system	Electrical components		
	Controlling part		
	Driving section		
Auxiliary part	Cooling system		
	Compressed air system		
	Lubricating system		

5.8 Part machine

Enter the CNC machining program and run it to complete the CNC machining of the part.

5.9 Part inspection

After the parts are processed, the workpiece should be carefully cleaned, and in accordance with the relevant requirements of quality management, the processed parts should be subject to relevant inspections to ensure the production quality. Table 5-9 shows the "three-level" inspection cards for machined parts.

Table 5-9 "Three-level" inspection cards for machined parts

Part drawing number		Part name		Working step number	
Material		Inspection date		Working step name	
Inspection items	Self-inspection result	Mutual inspection result		Professional inspection	Remark
Conclusion	☐ Qualified　☐ Unqualified　☐ Repair　☐ Concession to receive Inspection signature: Date:				
Non-conforming item description					

Project Summary

Through the CNC milling of the Geneva wheel, master the basic format of CNC milling program and the use of basic cutting instructions, and be able to use basic cutting instructions G1, G2, G3 to cut parts, and be able to flexibly apply coordinate rotation instructions G68, G69 and subprogram instructions M98, M99 to write processing programs.

Master the basic operation methods of vertical machining centers, including: powering on and off, cutter installation, workpiece alignment, cutter setting, program editing, graphic verification, CNC machining program debugging and automatic operation, etc.

Through task training, good professional quality, correct safety operation standards for machining centers, and basic machining quality awareness are developed and cultivated.

Exercises After Class

1. Fill in the blanks

(1) After the machine is powered on, first check if _____ is normal.

(2) In order to deal with the engineering drawing that uses the rotated partial view to represent the geometric size, we often use the method of _____ for CNC programming.

(3) There is no essential difference between the subprogram and the main program in the format, the main difference is _____ . The main program uses _____ to end the program, while the subprogram uses instruction M99 to end it and automatically return to the program that called it.

(4) The commonly used calling format of system subprogram is _____ .

(5) Check if there is any abnormality in _____ of the machine tool (such as the scatter shield, the footplate, etc.). Check if _____ of the machine tool are sufficient. Check whether there are foreign objects on the tool holder, the fixture, and the baffle plate of the lead rail. Check the state of each knob on the machine tool panel is normal. Check whether there is an alarm after power on the machine tool.

2. True or false

(1) The instruction to return the subprogram to the main program is instruction M98. ( )

(2) After processed, the processing burrs should be removed to ensure that the sharp corners are sufficiently blunt to ensure personal safety during use. ( )

(3) Usually, subprograms can be used as independent machining programs. ( )

(4) Subprograms can be called not only by the main program, but also by other subprograms. This method is called nesting of subprograms. ( )

(5) Use the instruction M98 to call the subprogram and specify the number of calls. The number of calls must be an integer. ( )

3. Choice questions

(1) The origin of the machine coordinate system is called ( ).
  A. workpiece zero point   B. programming zero point
  C. mechanical origin   D. space zero point

(2) The instruction to cancel the coordinate system rotation is ( ).
  A. G65   B. G66   C. G68   D. G69

(3) The subprogram return instruction is ( ).
  A. M96   B. M97   C. M98   D. M99

(4) When cutting grooved parts by the path method, the surfaces on both sides of the groove: ( ).
  A. one side is down milling, the other side is up milling

B. both sides are down milling

C. both sides are up milling

D. no need to do any processing

(5) When machining inner contour parts, (   ).

A. no need to leave a finishing allowance

B. in order to ensure down milling, the cutter should swing left and right along the surface of the workpiece

C. in order to ensure down milling, the cutter should move counterclockwise along the inner surface

D. The cutter should move in any direction along the workpiece surface

4. Short answer questions

Briefly describe the application of subprograms.

Self-learning Test Scoring Sheet shows in Table 5-10.

Table 5-10 Self-learning Test Scoring Sheet

Task	Task requirements	Score	Scoring rules	Score	Remark
Learn key knowledge points	(1) Master how to use the coordinate system rotation instructions (2) Master how to use subroutine instructions	20	Understand and master		
Technological preparation	(1) Ability to read part drawings correctly (2) Ability to independently determine the process technology and fill in the process documents correctly (3) Be able to write the correct processing program according to the processing process	30	Understand and master		
Hands-on training	(1) Be able to reasonably select gages according to the structural characteristics and accuracy of the parts, and correctly and normatively measure the relevant dimensions (2) Master the operation process of groove wheel milling (3) Be able to operate the CNC machining center correctly and adjust the processing parameters according to the processing situation	50	(1) Understand and master (2) Operation process		

Ideological and Political Classroom

Project 6 Programming and Machining Training for Threaded Hole of Drive Shaft and Driven Shaft

➢ Mind map

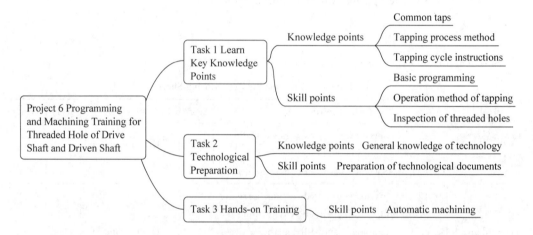

➢ Learning objectives

Knowledge objectives

(1) Know the working requirements of drive shaft and driven shaft.

(2) Know the use and key requirements of drive shaft and driven shaft in this mechanism.

Ability objectives

(1) Master the use of hole machining instructions.

(2) Master the use of threading instructions.

(3) Master the selection and use of threading cutters.

(4) Be able to independently determine the processing technology and correctly fill in the technological documents of threaded hole machining.

(5) Be able to operate the CNC lathe correctly and adjust the machining parameters according to the machining conditions.

(6) Be able to select measuring tools reasonably according to the structural characteristics and accuracy of parts, and be able to measure the relevant dimensions

correctly and normatively.

Literacy goals

(1) Cultivate students'scientific spirit and attitude.

(2) Cultivate students'engineering awareness.

(3) Develop students'teamwork skills.

Task 1   Learn Key Knowledge Points

6.1 Common taps

A tap is a tool for machining internal threads. According to the shape, it can be divided into spiral fluted tap, tool cutting edge inclination tap, straight fluted tap and pipe thread tap. According to the use environment, it can be divided into hand tap and machine tap. According to the specifications, it can be divided into metric, American, and British taps, etc. Taps are the most popular machining tools used by manufacturing operators when tapping.

Straight fluted taps are easy to process, with slightly lower precision and larger output. Generally used for thread processing of ordinary lathes, drilling machines and tapping machines, and the cutting speed is relatively slow. Spiral fluted taps are mostly used for drilling blind holes in CNC machining centers, with fast processing speed, high precision, good chip removal and good alignment. There is a chip holding groove in the front of the screw tap, which is used for processing through holes. Most of the taps provided by the tool factory are coated taps, which greatly improve the service life and cutting performance compared with uncoated taps. The tap with unequal diameter design has reasonable cutting load distribution, high machining quality, but high manufacturing cost. Trapezoidal thread taps are often designed with unequal diameters. Straight fluted taps and spiral fluted taps are shown in Figure 6-1.

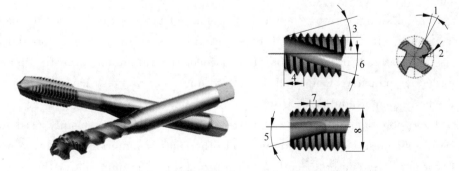

Figure 6-1 Straight fluted taps and spiral fluted taps

In addition to the nominal diameter and pitch of the tap (Shown in "8" and "7" in Figure 6-1), the hand tap and the machine tap should be known clearly when selecting the tap. The chamfer length (shown in 4 in Figure 6-1) can be used to identify the hand tap and the machine tap. Generally, the chamfer of the hand tap is long and that of the machine tap is short. In addition, pay attention to the tolerance. The commonly used tolerance zone numbers of international standard taps are ISO1, ISO2 and ISO3. ISO1 is used to process 4H and 5H threaded holes. ISO2 is used to process 5G and 6H threaded holes. ISO3 is used to process 6G, 7H and 7G threaded holes. Sometimes we can see some taps without tolerance zone remarked. Some materials define them as ISO4, which is used for hand taps to process 6H and 7H threaded holes.

6.2　Tapping process method

As shown in Figure 6-2, there are several common process methods for tapping.

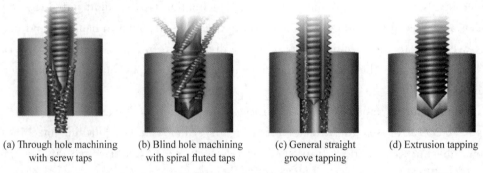

(a) Through hole machining with screw taps　　(b) Blind hole machining with spiral fluted taps　　(c) General straight groove tapping　　(d) Extrusion tapping

Figure 6-2　Tapping process

(1) Through hole machining with screw taps. This is a very powerful tapping process, suitable for harsh machining conditions. During the tapping process, the bottom hole is enlarged at the same time, and the chips are pushed forward through the hole.

(2) Blind hole machining with spiral fluted taps. This is a commonly used mechanical tapping technology. The spiral fluted tap is used in conjunction with the rigid tapping function of the machine tool, so that the chips can be discharged up along the spiral groove of the tap.

(3) General straight groove tapping. This kind of process can be used for hand and machine tapping. It can process all hole patterns, especially suitable for short chip materials (such as cast iron), and is often used in the automobile industry, pump and valve production industry. Chips fall downward due to gravity, so pay attention to chip removal when tapping blind holes from top to bottom.

(4) Extrusion tapping. It is a chip free tapping technology, mainly used for low carbon steel, stainless steel, aluminum, and other plastic materials. It can be used for various types of holes and hole depths. When processing aluminum alloy, the strength of its thread can be increased due to deformation strengthening. The taps used for extrusion tapping are often referred to as forming taps, and the corresponding other taps are cutting taps.

6.3 Thread bottom hole size for tapping cutting

When tapping the blind hole thread, the formula for calculating the depth of the bottom hole of the thread is:

$$H_{deep} = h_{effective} + 0.7 \times D$$

The diameter of the bottom hole is usually calculated by the following formula:

$$D_{Bottom\ hole} = D - P$$

where:

$h_{effective}$ is the effective depth of the thread.

D is the nominal diameter of the thread.

P is the thread pitch.

In practical applications, the bottom hole of the threaded blind hole can be machined slightly deeper. On the one hand, it prevents the tap tip from being stuck during tapping, and on the other hand, it can increase the chip space at the bottom. For different materials, the hole diameter is often slightly modified when machining threaded bottom holes. Generally, when machining brittle materials such as cast iron and bronze, 1.1-time pitch is deducted when calculating the aperture, and the 1-time pitch is deducted when machining plastic materials such as steel and copper. Therefore, these parameters should be flexibly selected according to the actual production, combined with the material and its heat treatment state.

6.4 Operation method of thread tapping

6.4.1 Hand tapping

The operation of hand tapping is shown in Figure 6-3.

(1) Install the tap into the tap hinge. Operation points: The tap hinge is used to clamp the tap to facilitate turning the tap for tapping. The tap hinge should be clamped on the square tenon of the taper shank, not on the smooth taper shank. Otherwise, there will be slippage between the hinge and the tap when tapping.

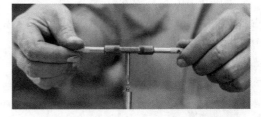

Figure 6-3 Hand tapping operation

(2) Insert the tap clamped on the hinge vertically into the bottom hole. Operation points: Cross-check the verticality of the tap and the port plane from the vertical and horizontal directions by visual inspection. Correct it if not vertical.

(3) Hold the hinge with both hands close together, and press down with your thumbs

against the middle of hinge. In a clockwise direction, press while turning, so that the tap gradually cuts into the hole. Operation points: Concentrate on the middle of the hinge with equal pressure, and try to make the tap cut into the hole vertically. The pressure should be appropriately larger, and the hinge should be turned slowly to prevent the port from slipping.

(4) When the tap is cut into the hole for 1~2 teeth, check the verticality of the tap. If skew is found, correct it. Operational points: Cross-check the verticality of the tap by visual inspection. If the teeth cut too far, force correction can damage the tap. Correction method: Slowly correct the deflection in the opposite direction while turning the hinge.

(5) After the tap is cut into the hole for 3~4 teeth, hold the hinge handle with both hands separately, no longer pressurize, and rotate the hinge handle evenly. For every 3/4 turn, rotate backward by 1/4 turn, and tap until it can't rotate. Lubricating oil should be added in an appropriate amount during the tapping process. Operation points: After tapping 3~4 teeth in the hole, some threads have been formed, just turn the hinge handle, do not pressurize, the tap will cut down by itself. If pressure tapping is continued, the formed thread will be damaged. Chips are formed during tapping, which will block the tap. The purpose of reverse rotation is to cut off chips and reduce resistance. Adding lubricating oil to reduce cutting resistance, which can reduce thread roughness and prolong tap life.

(6) Hold the hinge handle with both hands, turn it counterclockwise evenly and smoothly, and withdraw the tap from the hole. Clean the tap and the chips in the screw hole. Operation points: The hands should be evenly and smoothly inverted hinged. When the tap will be completely withdrawn from the hole, it shall be avoided that the tap shakes and damages the thread.

6.4.2 Mechanical tapping

Tapping is a common hole processing content on CNC machining centers. First, install the selected tap on a dedicated tapping pocket, preferably a floating pocket with stretch and compression features. The general steps of tapping are as follows.

Step 1: X, Y positioning.

Step 2: Select the spindle speed and direction of rotation.

Step 3: Quickly move to point R.

Step 4: Feed to the specified depth.

Step 5: Spindle stop.

Step 6: Spindle reverse rotation.

Step 7: Feeding back.

Step 8: Spindle stop.

Step 9: Quickly return to the initial position.

Step 10: Restart the normal rotation of the spindle.

Mechanical tapping is divided into flexible tapping and rigid tapping. Flexible tapping is mostly used in open-loop or semi-closed loop systems, the cost of which is much lower than that of closed-loop systems of rigid tapping. Flexible tapping has low processing efficiency due to accumulated errors and low spindle speed, and is suitable for small and medium batch production. Rigid tapping has a strict linear proportional relationship between the spindle speed and the feed. The selection of high-quality alloy tools can achieve high speed and high feed, so the processing efficiency is high, and it is suitable for mass production. Since there are few machine parameters involved in flexible tapping, the requirements for CNC debugging personnel are not high. In addition, the default state of the system is generally flexible tapping state. Rigid tapping involves many parameters, which is more complicated than flexible tapping.

In actual production, the machine tool cannot exactly match the pitch of the specific tap being used. There is always a slight difference between the thread produced by the machine and the actual pitch of the tap. If an integral tap holder is used, this difference has a decisive effect on tap life and thread quality because of the additional axial force on the tap.

If a tap clamp with tension compression floating is used, the tap life and thread quality will be greatly improved, because these additional axial forces on the tap are eliminated. For the traditional tension compression tap clamp, there is a problem that they will cause great changes in tapping depth. As the tap becomes dull, the pressure required to actuate the tap into the hole increases, and more compression stroke is used in the tap driver before the tap begins to cut, resulting in a shallower tapping depth.

One of the main advantages of rigid tapping is that the depth can be precisely controlled in blind hole machining. For precise and consistent machining of workpieces, tap holders with adequate compensation are required to achieve high tap life without causing any variation in depth control. For this training task, we choose rigid tapping for processing.

6.5 Tapping instructions

Tapping instructions G84 and G74 can be used for tapping in standard mode and rigid mode. Instructions G84 is used to process right-hand threads, and Instructions G74 is used to process left-hand threads. After executing the tapping instructions, the rotation or stop of the spindle of the machining center is controlled by the tapping instructions without human intervention.

Instructions G84 and G74 have the same parameters except for the thread direction of machining. Instructions G84 is taken as an example for introduction. Its format is as follows:

```
G84 X_ Y_ Z_ R_ P_ F_ K_ ;
```

Where, $X\_\_Y\_\_$ Threaded hole position;

$Z\_\_$ Distance from R point to the hole bottom or hole bottom position;

$R\_\_$ The starting position of tapping, which shall be in front of the formal tapping;

$P\_\_$ Hole bottom and pause time when returning to R point;

$F\_\_$ Cutting feed rate;

$K\_\_$ Number of repetitions (only when repetitions are required);

The action process is shown in Figure 6-4.

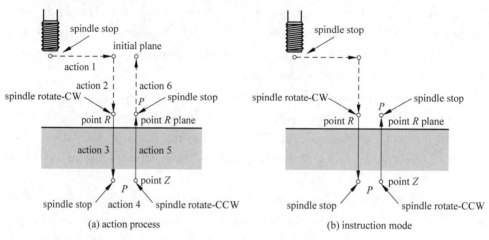

Figure 6-4 G84 tapping instruction action

The default mode is the flexible tapping mode. If rigid tapping is needed, it can be specified by the following three ways.

(1) Specify "M29 S__;" before tapping instruction.

(2) Specify "M29 S__;" in a block containing a tapping instruction.

(3) Specify the instruction as rigid tapping in the system parameters. For example, by setting the parameter G84 (No. 5200#0) to 1, the G84 instruction is used by default as rigid tapping.

Note that the feed override adjustment is invalid after entering the tapping instruction.

6.6 Inspection of threaded holes

Usually, threaded holes are measured with a thread plug gauge, which is shown in Figure 6-5. The thread plug gauge is a tool for measuring the correctness of the internal thread size. It can be divided into ordinary coarse thread, fine thread and pipe thread plug gauge. Thread plug gauge with a pitch of 0.35 mm or less and accuracy of class 2 or higher and thread plug gauge with a pitch of 0.8 mm or less and accuracy of class 3, both

Figure 6-5 Thread plug gauge

of them do not have end probes. Thread plug gauges below 100mm are taper shank thread plug gauges, and those above 100mm are double shank thread plug gauges. The thread plug gauge simulates the maximum solid thread form of the tested thread, and checks whether the effective pitch diameter of the tested thread exceeds the pitch diameter of its maximum solid thread form, and at the same time checks whether the actual size of the bottom diameter exceeds its maximum solid size by the "through end" and "stop end" of the thread plug gauge.

The thread plug gauges should not be operated violently. Usually, the go-gauge of the thread plug gauge can be rotated at any position of the tested thread, and it is judged to be qualified if it passes the entire length of the thread, otherwise it is an unqualified product. After the no-go gauge of the thread plug gauge is aligned with the measured thread, it is qualified if the screwed-in thread length is stopped between 2 pitches. It is not allowed to pass by force, otherwise it is judged as unqualified product.

Task 2   Technological Preparation

6.7   Part drawing analysis

As shown in Figure 6-6, according to the use requirements of the part, 45 steel is selected as the blank material threader hole of drive shaft, and the blank size is $\phi 50$ bar. Other parts have been processed on the CNC lathe before the threaded hole is processed.

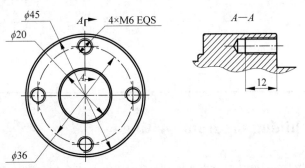

Figure 6-6   Part drawing of the threaded hole of drive shaft

The threaded hole in this process is a fastening connection hole, and the overall machining accuracy is not high. This training task is mainly to practice the basic tapping operation of the machining center, so it is relatively simple.

Note that this part is a manual operation part, so after processed, the machining burr should be removed to ensure that the acute angle is sufficiently blunt to ensure personal safety during use.

6.8 Technological design

According to the analysis of part drawing, determine the processing technology, as shown in Table 6-1.

Table 6-1 Working procedure card for milling

Machining working procedure card	Product model	CLJG-01	Part number	AXIS-01	Page 1
	Product name	Geneva mechanism	Part name	Drive wheel	Total 1 page

Procedure No.		
Procedure name	Hole machining	
Material	C45	
Equipment	Vertical machining center	
Equipment model	VAL6150e	
Fixture	Three-jaw self centering chuck	
Measuring tool	Thread plug gauge	
Preparation & conclusion time	60 min	
Single-piece time	10 min	

Work steps	Content	Cutter	S/ (r/min)	F/ (mm/r)	a_p/ mm	a_e/ mm	Step hours/min	
							Mechanical	Auxiliary
1	Workpiece installation							1
2	Machining center hole	Center drill	1500	100	0.05	1	1	
3	Drill $\phi 5$ bottom hole	$\phi 5$ twist drill	1200	100	0.1	2.5	3	
4	Chamfer	$\phi 10 \times 90°$	1200	100	0.1	2	1	
5	Rigid tapping	M6	600	60	1	0.8	2	
6	Disassemble and clean the workpiece							2

6.9 CNC machining program writing

According to the machining technology, the processing program is written as shown in Table 6-2. The drive shaft and the driven shaft are only different in cutter setting, and their CNC machining and process files are the same.

Table 6-2 CNC machining program of threaded hole of drive shaft and driven shaft

No.	Statement	Annotation
	O001;	Main program
N1	T1 M6;	Call the center drill
	S1500 M3;	
	G0 G90 G54 X18 Y0;	
	G43 Z10 H1;	

Continued

No.	Statement	Annotation
	G81 Z-1.5 R1 F100;	
	X0 Y18;	
	X-18 Y0;	
	X0 Y-18;	
	G80;	
	G0 Z100;	
	M5;	
N2	T2 M6;	Call $\phi 5$ twist drill
	S1200 M3;	
	G0 G90 G54 X18 Y0;	
	G43 Z10 H2;	
	G83 Z-17.7 Q2 R1 F100;	Use deep hole drilling instruction to process bottom hole with depth of 17.7mm (sharp point)
	X0 Y18;	
	X-18 Y0;	
	X0 Y-18;	
	G80;	
	G0 Z100;	
	M5;	
N3	T3 M6;	Call $\phi 10 \times 90°$ chamfer drill
	S1000 M3;	
	G0 G90 G54 X18 Y0;	
	G43 Z10 H3;	
	G81 Z-3.5 R1 F100;	
	X0 Y18;	
	X-18 Y0;	
	X0 Y-18;	
	G80;	
	G0 Z100;	
	M5;	
N4	T4 M6;	Call taps
	S600 M3;	
	G0 G90 G54 X18 Y0;	
	G43 Z10 H4;	
	M29 S600;	use rigid tapping instruction
	G84 Z-16 R1 F60;	When setting the tapping depth, comprehensively consider the tip of the tap to ensure that the effective thread is 10mm
	X0 Y18;	
	X-18 Y0;	
	X0 Y-18;	
	G80;	
	G0 Z100;	
	M5;	
	M30;	

Task 3　Hands-on Training

6.10　Equipment and appliances

Equipment: AVL650e vertical machining center.

Cutters: $\phi 3$ center drill, $\phi 5$ twist drill, $\phi 10 \times 90°$ chamfer drill, M16 machine tap.

Fixture: the K11-320 three-jaw self-centering chuck.

Tools: kry file.

Measuring tools: 0~150mm vernier caliper, 0.02mm lever arm test indicator (with mounting rod), M6-6H thread plug gauge.

Blank: $\phi 45$ (after turning).

Auxiliary appliances: Chuck wrench, rubber hammer, brush, etc.

6.11　Check before powering on

Check whether there is any abnormality in various parts of the appearance of the machine tool (such as the scatter shield, the footplate, etc.). Check whether the lubricating oil and coolant of the machine tool are sufficient. Check whether there are foreign objects on the tool holder, the fixture, and the baffle plate of the lead rail. Check the state of each knob on the machine tool panel is normal. Check if there is an alarm after power on the machine tool. Refer to Table 6-3 to check the machine state.

Table 6-3　Preparing card for machine start-up

Check item		Test result	Abnormal description
Mechanical part	Spindle		
	Feed part		
	Tool holder		
	three-jaw self centering chuck		
Electrical part	Main power supply		
	Cooling fan		
CNC system	Electrical components		
	Controlling part		
	Driving section		
Auxiliary part	Cooling system		
	Compressed air system		
	Lubricating system		

6.12 Part machine

Enter the CNC machining program and run it to complete the CNC machining of the part.

6.13 Part inspection

After the parts are processed, the workpiece should be carefully cleaned, and in accordance with the relevant requirements of quality management, the processed parts should be subject to relevant inspections to ensure the production quality. Table 6-4 shows the "three-level" inspection cards for machined parts.

Table 6-4 "Three-level" inspection cards for machined parts

Part drawing number		Part name		Working step number	
Material		Inspection date		Working step name	
Inspection items	Self-inspection result	Mutual inspection result	Professional inspection		Remark
Conclusion	☐ Qualified ☐ Unqualified ☐ Repair ☐ Concession to receive Inspection signature: Date:				
Non-conforming item description					

Project Summary

Through the threading of drive shaft and driven shaft, master the basic format of threading program and the use of basic cutting instructions, and be able to use basic threading instructions G84 and instructions G74 to tap the thread hole, and be able to flexibly apply M29 instruction to write rigid tapping program.

Master the basic operation methods of vertical machining centers, including: powering on and off, cutter installation, workpiece alignment, cutter setting, program editing, graphic verification, CNC machining program debugging and automatic operation, etc.

Through task training, good professional quality, correct safety operation standards for machining centers, and basic machining quality awareness are developed and cultivated.

Exercises After Class

1. Fill in the blanks

(1) Common tapping processing methods include _____, _____, _____ and _____.

(2) In the formula for calculating the depth of the threaded bottom hole, h represents _____, D represents _____, and P represents _____.

(3) For different materials, the hole diameter is often slightly modified when machining threaded bottom holes. Generally, when machining brittle materials such as cast iron and bronze, subtract _____ times the pitch when calculating the hole diameter, and subtract _____ times the pitch when machining plastic materials such as steel and copper.

(4) The operation methods of thread tapping include hand tapping and _____.

(5) In the tapping instructions G84, the meaning of K is _____.

2. True or false

(1) Straight fluted taps are easy to process, with slightly lower precision and larger output. Generally used for thread processing of ordinary lathes, drilling machines and tapping machines, and the cutting speed is relatively slow. ( )

(2) General straight groove tapping, chips fall down due to gravity, so pay attention to chip removal when tapping blind holes from top to bottom. ( )

(3) When hand tapping, add lubricating oil and evenly turn the hinge, the tap will cut down by itself, just need to tap until it stops turning. ( )

(4) In actual production, the machine tool cannot exactly match the specific tap pitch being used. There is always a slight difference between the thread produced by the machine and the actual pitch of the tap. ( )

(5) Instructions G84 and G74 have the same parameters except for the thread direction of machining. Instructions G84 is used to process left-hand threads, and instructions G74 is used to process right-hand threads. ( )

3. Choice questions

(1) When processing hole parts, the method of drilling→flat bottom drilling and reaming→chamfering→fine boring is applicable to ( ).

    A. stepped hole   B. blind hole with small diameter

    C. blind hole with large diameter   D. Flat bottom hole with large diameter

(2) Mechanical tapping is divided into flexible tapping and rigid tapping. The wrong one of the following options is ( ).

    A. Flexible tapping is mostly used in open-loop or semi-closed loop system, the cost of which is much lower than that of closed-loop systems of rigid tapping.

B. Because of the accumulated error and low spindle speed, flexible tapping has low processing efficiency and is suitable for small and medium batch production.

C. Rigid tapping has a strict linear proportional relationship between the spindle speed and feed, and it adopts high-quality alloy cutters to realize high speed and high feed, so the processing efficiency of rigid tapping is high, and it is suitable for mass production.

D. The system is generally in a rigid tapping state by default.

(3) G84 tapping instruction action can be divided into (   ) parts in G98 mode.

A. 5　　　　　　B. 6　　　　　　C. 7　　　　　　D. 8

(4) The following is not a thread plug gauge is (   ).

A. ordinary coarse thread　　　　B. fine thread

C. round hole plug gauge　　　　D. pipe thread

(5) Which of the following options is not used for thread machining. (   )

A. G32　　　　　B. G92　　　　　C. G76　　　　　D. G81

4. Short answer questions

(1) Briefly describe the types of taps.

(2) Briefly describe the functions and differences of instructions G74 and G84.

(3) Briefly describe the detecting method of the thread.

Self-learning test scoring sheet is shown in Table 6-5.

Table 6-5　Self-learning Test Scoring Sheet

Task	Task requirements	Score	Scoring rules	Score	Remark
Learn key knowledge points	(1) Learn about the classification of common taps and their applications (2) Master several common process methods of tapping processing (3) Master the operation method of thread tapping (4) Master the use of tapping instructions G84 and G74 and understand the meaning of each parameter (5) Master the correct measurement method for threaded holes	25	Understand and master		

Continued

Task	Task requirements	Score	Scoring rules	Score	Remark
Technological preparation	(1) Ability to read part drawings correctly (2) Ability to independently determine the process technology and correctly fill in the threaded hole processing process documents (3) Be able to write the correct processing program according to the processing technology	25	Understand and master		
Hands-on training	(1) It can reasonably select gages according to the structural characteristics and accuracy of parts, and correctly and normatively measure the relevant dimensions (2) Master the selection and use of threading tools (3) Master the operation process of drive shaft and driven shaft threaded hole processing (4) Be able to operate the CNC machining center correctly and adjust the processing parameters according to the processing situation	50	(1) Understand and master (2) Operation process		

Ideological and Political Classroom